The Psychology of Consciousness: Theory and Practice

Hashim Talib Hashim • Athanasios Alexiou
Editors

The Psychology of Consciousness: Theory and Practice

Springer

Editors
Hashim Talib Hashim
University of Baghdad - College of Medicine
Baghdad, Iraq

Athanasios Alexiou
Novel Global Community Education Foundation
Hebersham, NSW, Australia

ISBN 978-3-030-90691-7 ISBN 978-3-030-90692-4 (eBook)
https://doi.org/10.1007/978-3-030-90692-4

This Springer imprint is published by the registered company Springer Nature Switzerland AG
The registered company address is: Gewerbestrasse 11, 6330 Cham, Switzerland

I want to dedicate this work to my father (Talib Hashim Manea) who supported and cared for me all these years.

To my mother (Jawaher Mutar Mohammed) for everything she has given me since I was born till now.

To my grandmothers and grandfathers who have given us such a great and inspiring family.

To my brothers and sister who never give up on me, and they were there all the time to support me.

To my friends and to everyone said a supporting word one day that made me continue in our way.

To my professors and to my teachers and to everyone who encouraged me to do it, especially Dr. Samer Hoz who inspired me to write and edit this book.

Foreword

The Psychology of Consciousness: Theory and Practice is a fundamentally important book about one of the most important topics for the twenty-first century. It combines the best of psychology, neuroscience, and consciousness studies, as well as requires other disciplines to add voice to the discussion including physics, chemistry, and biology. Consciousness itself, of course, is the most mysterious of all things that we contend with as human beings. We all feel quite confident that we have consciousness and yet, it seems that it always eludes discovery whenever we search for it with our scientific tools. Part of the reason for this problem is the consciousness by its very nature is subjective. We exist within our consciousness, and so do all of our investigations of consciousness. If science, psychology, and all the other disciplines used to understand consciousness exist within our own consciousness, we are at an immense disadvantage in trying to discover exactly what consciousness is. But that is the lofty goal of this book which strives to understand the psychology of consciousness with great breadth and depth of investigation.

This book brings together all of the new studies that have been exploring consciousness and combines that understanding with ancient perspectives, spiritual perspectives, and certainly multidisciplinary perspectives. This latter point is perhaps the most important. The investigation of consciousness absolutely requires a complex and integrative approach blending many fields of study and scholarship. A related field of my own research focuses on neurotheology which seeks to understand the integrative relationship between the brain and our religious and spiritual selves. There is interesting overlap between neurotheology and the study of consciousness because so many intense spiritual or mystical states are associated specifically with an altered state of consciousness. And vice versa, altered states of consciousness are often described as being religious or spiritual. As a simple example, recent research in the field of psychedelics has demonstrated that intense experiences under the influence of drugs such as psilocybin are often reported to be the most spiritual of all experiences for the individual. So, there is certainly a complex relationship between intense spiritual experiences and alterations in the sense of consciousness. Data such as these are part and parcel of a broader exploration of the psychology of consciousness. And there are many other fascinating examples from normal states of consciousness that are experienced by most people to pathological states of the brain. Importantly, the authors do not shy away from a full exploration of the topic of the psychology of consciousness.

They appropriately begin with definitions of consciousness. While this seems to be a natural starting point, in and of itself, it is no easy task. First, consciousness by its very nature is difficult to define. We can start with historical definitions, but we must include more recent work to more deeply understand what consciousness is. It is also interesting to consider consciousness from various scholars including philosophers, theologians, biologists, and even those who pursue artificial intelligence. Each of these different perspectives may provide a slightly different type of definition for consciousness—some more practical and some more esoteric. In addition to trying to drive a definition of consciousness, we also have to consider a more operationalized perspective. In other words, how does consciousness manifest itself within the brain, the mind, and perhaps even the soul. These interrelationships are also essential for understanding consciousness and also for pursuing future explorations both theoretical and practical. Perhaps most important in the discussion of what consciousness is relates to the relationship between the brain and the mind. The age-old question between the dualistic approach that separates the brain from consciousness or an integrative approach that combines them in some way is a fundamental target for this book. If the brain creates consciousness, that leads to an entirely different kind of approach than if consciousness creates the brain. These are the fascinating questions that are explored in the pages that follow.

One of the troublesome problems of consciousness is not just a basic definition but the recognition of different levels of consciousness. There is the unconscious mind that influences us, without us ever knowing it. Preconsciousness and subconsciousness have an influence on our awareness. There is the conscious mind which is the primary target of our understanding. But there is even a post-conscious mind that helps us look back on our own awareness. Each of these needs to be explored in great detail and is tackled by the authors of *The Psychology of Consciousness*.

Altered states of consciousness are an essential target of study when it comes to understanding the larger concept of consciousness itself. They are also helpful in bringing these questions to a more practical realm. By this I mean that everyone has experienced certain states of consciousness throughout their everyday lives. The most basic state of consciousness is waking consciousness which we use when we go to work or school, talk with a friend, or watch television at night. This state of consciousness is contrasted by the other most well-known state of consciousness—sleep. We have all experienced our sleep consciousness and the different elements that arise within it, particularly dream states. Dream states can be rather chaotic and brief or long and complex. It is also interesting to note that when one shift from one state of consciousness to another, we take different perspectives of those states. By this I mean that no matter how real a dream consciousness state may feel, when you wake up, you immediately relegate that state to an inferior state of consciousness or reality. This latter term is also an important part of understanding consciousness since it is through our consciousness that we establish what reality actually is. But what do we make of altered states of consciousness such as dream states, drug-induced states, or even mystical states when it comes to which is real? It seems that each of these states of consciousness carry with them a certain level of "realness," which we are able to compare against each other. But no matter how real any of these states feel, they are all housed within our consciousness.

When considering these different states of consciousness, we also have to tie everything back to our psychology and our biology. When talking about a drug-induced state, we can ask the question as to what parts of the brain are affected by the drug? Many psychedelics affect the serotonin system, for example. So why does intense stimulation of the serotonin system lead to these powerful experiences and altered states of consciousness. Is consciousness in the serotonin molecule itself? Or does serotonin activate the neurons of the brain in such a way that facilitates these shifts in consciousness? And what can we say about psychopathological states such as depression or schizophrenia in the context of consciousness? Is there a way to explore consciousness through the lens of psychopathology, and perhaps even turn that discussion around to help those with these pathologies? The answers to such questions drive at the heart of what consciousness actually is and how it works with respect to the brain.

Moving beyond consciousness associated with human beings and perhaps other animals, consciousness is within the field of computer-based artificial intelligence. There has been a long debate as to whether we can produce a computer that has consciousness. The argument focuses on whether consciousness is derivative of a system's degree of complexity or whether there is something beyond a complex set of connections that leads to consciousness. If consciousness exists within the brain simply because of its complex interconnected network of neurons, then arguably, we can produce a complex enough computer that would become conscious itself. But if consciousness derives from something more fundamental, it may be found only in biological organisms. Alternatively, if consciousness is something that exists beyond all physical matter, then it can potentially manifest in any physical system. These are the types of fascinating questions addressed in the pages to follow.

In those pages, the authors will consider a great many of new findings related to consciousness and strive to develop an understanding that provides important information for many of the interrelated fields. Thus, this information is important for the neuroscientist, the psychologist, the philosopher, and anyone else in the field of consciousness studies. This book encapsulates a great many ideas and provides a background that will help people to understand this incredibly complex topic. The authors are also not afraid to explore a variety of consciousness-related topics that may not be as popular as current mainstream ideas. But given the complexity and the challenges around the understanding of consciousness, it is always important to be open to the many possibilities, even those that seem unlikely. The famous author and futurist, Arthur C. Clarke, once said, "The only way of discovering the limits of the possible is to venture a little past them into the impossible." This book provides a little bit of a look at both the possible and the impossible and provides an essential reading for anyone who wants to learn more about this curious thing called consciousness.

Andrew Newberg
Department of Integrative Medicine
and Nutritional Sciences
Marcus Institute of Integrative Health,
Thomas Jefferson University, Philadelphia, PA, USA

Department of Radiology
Marcus Institute of Integrative Health,
Thomas Jefferson University, Philadelphia, PA, USA
7 December 2020

Foreword

The concept of consciousness has many meanings that are used depending on the scientific discipline dealing with this idea. The term is also used in everyday speech. We can describe a man who can critically evaluate his behavior as a person who is conscious of his behavior. The same is stated by court experts when examining a criminal who has premeditatedly committed atrocious acts. A conscious choice is deciding about an activity after considering the benefits and consequences of such behavior. Conscious decisions have been confirmed in numerous studies by parents adopting a child, transsexuals who decide to correct their gender, or as is the case in some countries, chronically ill people who decide to consciously take their own lives (euthanasia). Consciousness is therefore intentional (Husserl 2002). It is a certain type of knowledge about a given topic that leads to making a deliberate choice.

One could also say that consciousness has its scope. We speak of global consciousness in the context of targeted measures to improve the quality of life on our planet (Elgin 1997).

Environmental consciousness is the understanding of the relationship between human activities and the world of nature (Bonnett 2017). Social awareness is the totality of ideas, opinions, beliefs, and views—the way we understand and define the environment—the totality of scientific, philosophical, ideological, political, legal, religious, and ethical views that make up the way of thinking and culture of a given society (Durkheim 2003). Awareness of one's surroundings allows you to move safely in a certain area or avoid a place due to the hazards there. Self-consciousness is the ability to undertake activity adequate to the circumstances, using cultural capital (collective consciousness), learned skills, and respect for values and principles.

Consciousness may be disturbed or lost as a result of a sudden event or deterioration of health. It is also possible not to be aware of something due to the lack of knowledge in a given area or because of strong emotions that interfere with a rational assessment of the situation. Consciousness can be acquired, and by planning or unintentional actions, one can also deprive someone of awareness. Consciousness can be deepened, expanded, enriched, and impoverished. Consciousness is therefore plastic.

Consciousness can also be manipulated. Altered states of consciousness are most often deliberate modifications of the perception of oneself and the environment, provoked by means of having such potential (Bayne and Carter 2018). While the incidental use of mood-altering agents should rather not leave permanent traces in consciousness, frequent or habitual use of these agents may lead to profound disturbances of self-awareness (the conscious self), requiring specialized long-term psychotherapy.

Pure consciousness is a state of a deep and alert mind (Husserl 2002). Dirty consciousness is knowledge about oneself and one's behavior distorted by defense mechanisms (Woydyłło 2014). Researchers of addicts indicate several forms of distortion of self-awareness caused by the abuse of intoxicants in this group: dispersion, splitting, or even hollowing out the structure of the "I" (Mellibruda and Sobolewska-Mellibruda 2006).

In recent decades, the concept of consciousness has been enriched with another designate. When speaking of conscious machines, we mean extremely efficient computer systems that can perform work just like a human, and even almost completely exclude human participation. It is true that calling these machines conscious is subject to numerous discussions, but researchers indicate that technical progress does not exclude such achievements in the future (Al-Imam et al. 2020).

The above examples of defining consciousness are only a modest fragment of what many researchers propose, and the book that the reader holds in his hands skillfully combines previous achievements with proposals for further expanding knowledge in this field.

Consciousness and its disorders are subjects close to my profession. As an addiction therapist, in every case of working with a patient, I encounter at least a double image of what is fact and what has been grounded by distortions of the perception of the person who is going to be treated. In addicts, disturbances of consciousness, especially self-consciousness and one's abilities, are a consequence of dualistic self-perception in situations related to being under the influence of intoxicants and being sober. Blocking is a common procedure triggered by the addict's own defense mechanisms (especially by the mechanisms of illusion and denial). A strong desire to forget or reject unpleasant events can block the access of consciousness to unwanted and unpleasant memories for many years. Such displacement from memory may in some way protect against unacceptable facts from one's own life by not allowing one's behavior to change. However, displaced events can survive on the unconscious level and cause emotional conflicts for a long time. The mechanism of denial created and strengthened during the development of addiction is an effective barrier against the acceptance of any information indicating the negative consequences of using mood-altering agents. The "disarmament" of this mechanism and unblocking of the consciousness during therapy allows us to understand that past failures and errors are the result of using these agents. Withdrawal of drugs or alcohol will not erase an unfavorable image from memory, but it can certainly stop the process of its distortion and may initiate the accumulation of new, better memories (Motyka 2015). Self-consciousness can therefore be restored, and not only to those who lost it as a result of illness or accident, but also to those who had manipulated it. This is just one example of the possibilities of working on consciousness.

Learning about consciousness is a journey. Cognition may be an aim in itself, but there may be many ways to obtain this knowledge. The traveler may choose a direction determined by philosophy, psychology, neurobiology, or other sciences referring to processes related to the creation, acquisition, consolidation, or change of consciousness. They can also try to cover each of these domains.

It was an honor for me to write a foreword to this book. Having read the chapters that were sent to me, I am proud to share my thoughts and encourage you to read this book.

As I wrote above, learning about consciousness is a journey. Therefore, I invite readers to the extraordinary journey that this book is. Each chapter can be a supplement to the others, and all of them can certainly contribute to the expansion of our knowledge (and therefore consciousness as well). The journey begins with an introduction to the understanding of consciousness from ancient cultures, through the deliberations of philosophers and psychoanalysts, to the contemporary definition of consciousness and an interesting theory of post-consciousness proposed by the authors. Colorful illustrations, similar to beautiful landscapes, allow you to stop during this journey and record the adventure with this book better. The continuation of the educational journey is getting to know the relationship between the brain and the mind, exploring the essence of the levels of consciousness, both human and machine. The next stages of the journey through the land of consciousness are other interesting places, the exploration of which will certainly result in the acquisition of new knowledge that can be stored in our personal "storage room." Hypnosis and awareness, sleep and dreaming, defense mechanisms and altered states of consciousness are areas that are definitely worth exploring. The chapter describing changes in consciousness caused by the agents causing such modifications is a particularly interesting stage in my adventure with this book. This is a difficult but definitely interesting subject. It is good that it was included in the "journey" proposed by the authors.

The book is aimed at doctors, psychologists, psychiatrists, philosophers, and mental health professionals, but in my opinion, representatives of other scientific disciplines can also use it. Importantly, the accessible narrative makes this book suitable for students at all levels of education.

The authors deserve special recognition for their holistic approach to the subject matter, for taking up a topic that is not easy, yet presenting it in an accessible way while adding something new in this regard. Each chapter encourages you to continue this intellectual and scientific journey and move on to the next pages. There are no boring clichés and unnecessary duplications of what has already been done in this area. Instead, there is good scientific text that is fresh, evokes reflections, and remains in the reader's memory.

It is a pleasant educational journey.

Let us travel then!

Marek A. Motyka,
University of Rzeszów
Rzeszów, Poland
11 December 2020

References

Al-Imam A, Motyka MA, Sahai A, Konuri VK. The "March of Progress": from cosmic singularity to digital singularity. Curr Trends Inf Technol. 2020;10(1):1–8.

Bayne T, Carter O. Dimensions of consciousness and the psychedelic state. Neurosci Conscious. 2018;1:niy008. https://doi.org/10.1093/nc/niy008.

Bonnett M. Environmental consciousness, sustainability, and the character of philosophy of education. Stud Philos Educ. 2017;36:333–47.

Durkheim E. Les formes elementaires de la vie religieuse Le systeme totemique en Australie [The elementary forms of the religious life]. Paris: Presses Universitaires de France; 2003.

Elgin D. Global consciousness change: indicators of an emerging paradigm. 1997. https://duaneelgin.com/wp-content/uploads/2010/11/global_consciousness.pdf. Accessed 11 Dec 2020.

Husserl E. Ideen zu einer reinen Phänomenologie und phänomenologischen Philosophie [Ideas pertaining to a pure phenomenology and to a phenomenological philosophy]. Berlin: De Gruyter; 2002.

Mellibruda J, Sobolewska-Mellibruda Z. Integracyjna psychoterapia uzależnień [Integrative addiction psychotherapy]. Warszawa: Instytut Psychologii Zdrowia; 2006.

Motyka M. Pamięć niszcząca, pamięć uzdrawiająca: labirynty wspomnień osób uzależnionych od środków psychoaktywnych [Destructive memory, healing memory: labyrinths of memories of people addicted to psychoactive substances]. Problemy Higieny i Epidemiologii. 2015;2(96):321–8.

Woydyłło E. Dobra pamięć, zła pamięć [Good memory, bad memory]. Kraków: Wydawnictwo Literackie; 2014.

Foreword

I believe something magical can happen when you read books.

I believe that one day the girls I see with books
all the time become great writers.

Indeed, a moment of pride.

When passion meets profession and you surround yourself with people having the same passion as yours, motivating each other, they end up doing what their heart says and not what people say: that's when history is created. This is the success mantra of many great people "Follow your dreams." Today, their dream of writing has turned to reality in the form of a book and has reached your hands which is now a part of history.

The aim of this book is to acquaint the readers with the multidimensional aspects of human psychology in a much easier way.

The content has been woven into short precise interesting parts to make reading of the most complicated subjects easy to understand.

Interestingly, there has been a dearth of literature on this complicated subject. It was my dream too to write a book on my most favorite subject "The Human Psychology" to reach out to international readers. These future doctors have made my dream come true.

I am proud of them, and they have it in them to power on.

More specifically, the objective of this book is to provide breadth and insight into the subject with the view to stimulate the readers, so that they involve themselves actively while reading and understanding the concepts of human psychology well.

I acknowledge the efforts of the young team behind this book.

Their hard work is appreciated, and I wish them all the best.

I hope the readers will find this book useful.

Lokesh Babu,
Sneha Mano Vikasa Kendra,
Sejal New Life Foundation
Bangalore, India

Preface

The consciousness is something not well understood, and it is vague in many aspects, so trying to understand it well is something impossible, but many theories have emerged to discuss it. And it is necessary to write an independent book about it to make it a popular subject for research and studies.

This book talks about the levels of consciousness and their roles in controlling our life and behavior. He consciousness ha a main role in learning human to behave and to live in all life's situation and ages. This book clarifies these situations in detail and the laws that make this system work properly. It provides many solutions and suggestions to control ourselves and our minds and put them in the right way. This book explains many of our behaviors depending on the psychology and the role of the consciousness in psychiatry, how to treat diseases and mental disorders, and how to improve the mental health as well.

This subject is not well discussed and detailed in literature so there is a need to give this topic its role in the psychology and in scientific literature too.

This book targets the consciousness' levels and the role of these levels in our life and behavior, so it divides the roles among them as appropriate and in the right way and then the humans can recognize which part is more important than the other and on what they should focus.

We have tried to put all the knowledge related to consciousness in this book from A to Z by making one understand the modern theories and studies clearly.

What makes this book unique is collecting all the theories and explanations of the consciousness in an integrated style so every chapter is related to the other and completes it.

We hope that the students, researchers, professors, and readers will understand the consciousness in a very easy and understandable way that allows everyone in any level to have an idea about consciousness.

Baghdad, Iraq — Hashim Talib Hashim
Hebersham, NSW, Australia — Athanasios Alexiou

Acknowledgments

We want to thank those people who helped us during the journey of writing this book. They are:

Mustafa Ahmed Muhson, Yehya Zia Tuama, Ahmed Dhiya Sagban, Shoaib ahmed, Mohammed Yasir Essar, Dr. Samer Hoz, Christos Tsagkaris, Ali Mahdi Mansour, Mojtaba Talib, Mohammed Qasim Mohammed, Ahed El Abed El Rassoul, Abbas Kamil, Abbas Mutar, Mohammed Abbas, Mariam Alaa, Fatima Abbas, Dr. Nabeel Al Khateeb, Sarmed Kadhem, Zain AL Saleh, Intsar Jabbar, Dr. Alaa Jameel, Dr. Hussein Al Zaidy, John Bchara, Morad Yasir Al Mostafa, Abdallah Reda, Nazmus Sakib Choudhary, Fahtiha Nasreen, Hadi Hussein, and Muslim Hassan Manea.

Contents

Introduction 1
Hashim Talib Hashim, Mustafa Ahmed Ramadhan, and Mehek Cheema

Brain and Mind 19
Hashim Talib Hashim and Adil Alhaideri

Levels of Consciousness 27
Hashim Talib Hashim and Mustafa Ahmed Ramadhan

Sleep and Dreaming 39
Saleh Abdulkareem Saleh

Defense Mechanisms and Personality Disorders 57
Saleh Abdulkareem Saleh

Drugs and Consciousness 81
Mustafa Hayder Kadhim

The Altered States of Consciousness 95
Mustafa Hayder Kadhim

Hypnosis and Consciousness 109
Hashim Talib Hashim and Mustafa Ahmed Ramadhan

Artificial Consciousness 119
Hashim Talib Hashim and Mustafa Ahmed Ramadhan

Relativity of the Human Mind 127
Hashim Talib Hashim and Mustafa Ahmed Ramadhan

Death and Consciousness 131
Hashim Talib Hashim and Mustafa Ahmed Ramadhan

Index 135

Introduction

Hashim Talib Hashim, Mustafa Ahmed Ramadhan, and Mehek Cheema

1 Mayan and Incan Theories of Consciousness

These are the early recognized explanations of consciousness. Despite being primitive, they created a steppingstone for developing consciousness theory (Wilber 2004).

They suggested a sensible organization, a sensible organization, a temporal relationship between the consciousness levels and the humans. "The awareness of being aware" is the definition used by Shamans and priests (Mayan Theories) (Fig. 1).

It is recognized as a branch of metacognition (higher order thinking skills, defined as cognition about cognition, thinking about thinking, and knowing about knowing). Thus, for example, these levels will be organized in a pyramidal shape from the cellular level to a universal level as shown in Fig. 2.

These are the steps of the development of consciousness according to Mayan theories. At the same time, Incan theories related to civility, which means progressing from level to level in relation to other changes and concerns (Fig. 3). Thus, the third level is recognized as the level of most people, and the fifth level is the sign of Taripay Pacha (Inca day of judgment).

H. T. Hashim (✉) · M. A. Ramadhan
University of Baghdad, College of Medicine, Baghdad, Iraq

M. Cheema
Vardhman Mahavir Medical College, New Delhi, India

H. T. Hashim, A. Alexiou (eds.), *The Psychology of Consciousness: Theory and Practice*, https://doi.org/10.1007/978-3-030-90692-4_1

Fig. 1 The first definition of consciousness

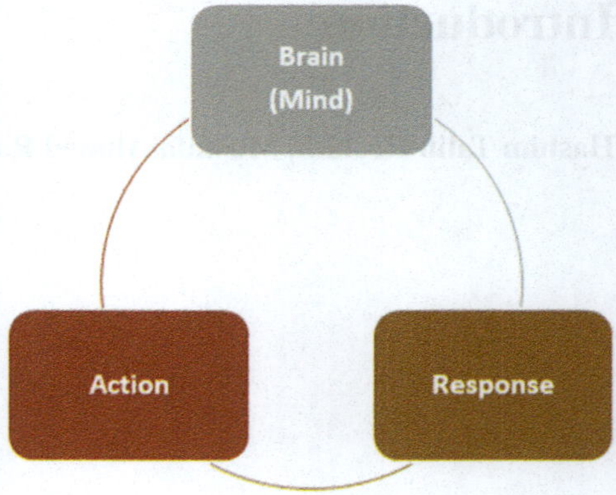

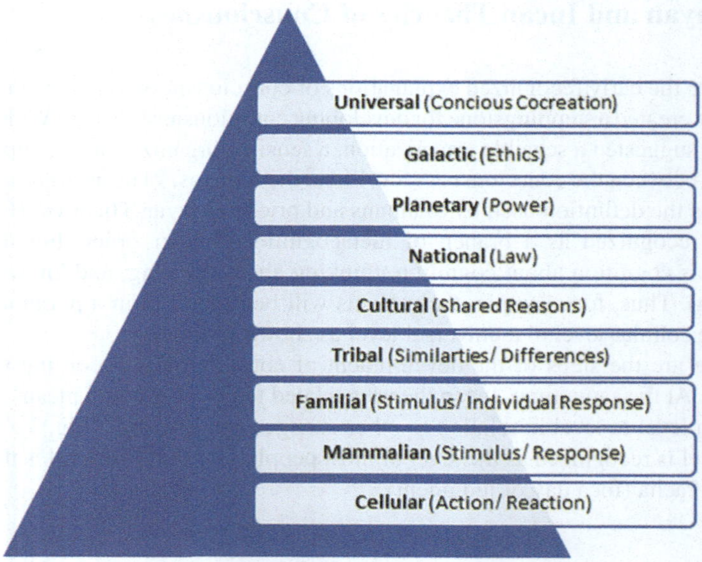

Fig. 2 The steps of the development of consciousness (Pons and Harris 2001; Morin 2006)

2 John Locke on Consciousness

The philosopher, John Locke, was from the earliest interested in consciousness. The main idea of his theory was "Personal ID" and the "Consciousness after death." He thought that the personal human ID is essential for psychology's continuity. He considered personal ID or the human's self as part of consciousness, but it is not involved in the soul or the body matters (Kriegel 2006).

Personal identity theory is a confrontation in philosophy with the questions humans ask about their origin and presence and the end of their life, like who is the

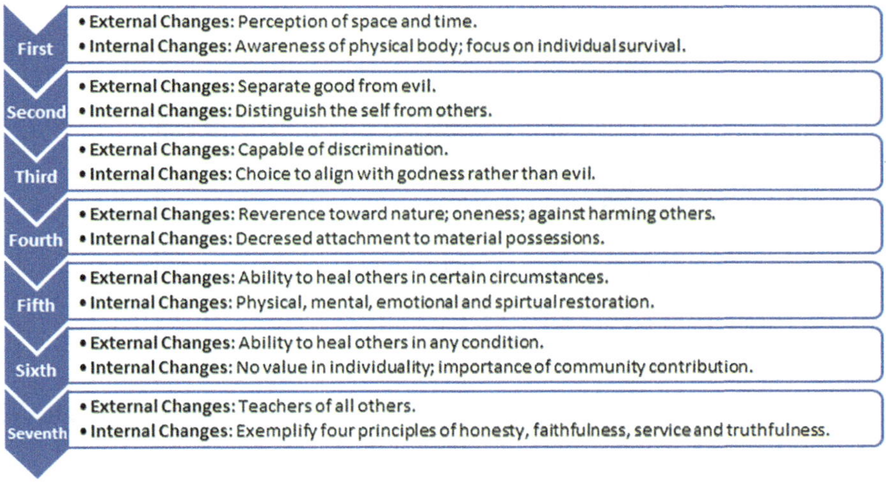

Fig. 3 The development of consciousness according to the Incan theories

human, why we are here, will live another life after we die, etc. So, it aligns with consciousness, starting with it and ending with it consequently.

According to Locke's theory, "Understanding is one of the first model conceptualizations of consciousness as repeated self–identification" (Nimbalkar 2011). Locke created what'' called an "Empty Mind" full of nothing, which can be created by ideas such as the experience, the sensations, and the reflections. He created a common concept between the body and the soul, which might not be acceptable to those who identify consciousness as the brain. According to his theory, the brain is similar to the body and, similarly to any other substance, may alter while consciousness remains identical. Therefore, personal identity cannot be located in the brain but within consciousness. The issue with this theory is life after death because to stay alive after death, there should be a person similar to the person who was there before death.

He suggested that consciousness can move from anyone's soul to another one's soul, and he thought that the Id moves with it.

John thought that God sees differently from others, so people will judge you about what they see, but God does not. As a result, the human will be responsible for the actions for which the human is conscious (Butler 1975; Reid 1785).

John concluded that the actions and ideas that the person makes during youth would shape the future and be the cornerstone for the upcoming actions.

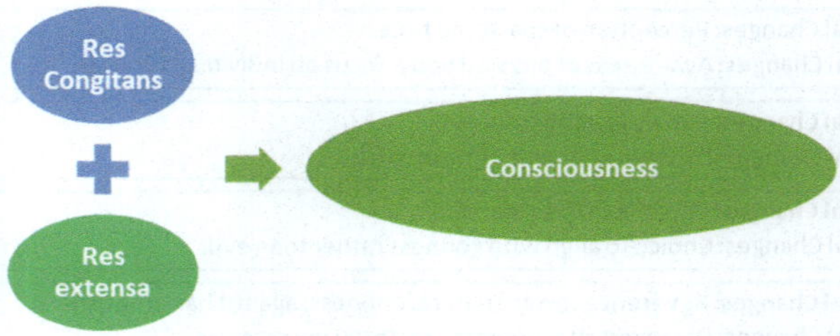

Fig. 4 The two domains of consciousness according to the Cartesian Dualism

3 René Descartes on Consciousness

It discussed Cartesian Dualism, which concludes the coming of nonphysical thing, which is the consciousness from the physical thing, the human body. This theory has two domains that consciousness resides in, which are given in Fig. 4:

1. ***Res Cogitans (The realm of thought) is the immaterial domain.***
2. ***Res Extensa (The realm of extension) is the material domain.***

The correlation of mind and body is not about how they can react but beyond that. It is about how they can be related and how they affect one another. The characteristics of each one is different from the other.

The mind does not occupy a space as the body does. The theory of Descartes suggests that consciousness is not a matter like the brain. It has no shape or physical properties. Human bodies indeed occupy space, and minds do not, in the very straightforward sense, assign linear dimensions and locations to the mind, or its contents and activities are obscure (Miller 2005).

Mind and body interactions are impossible because it contains contraindication of in space existence, so he suggested that the nonphysical (Not in Space) must act on what is physical (in space). He claimed a more robust and more immense power outside the human consciousness that affects the mind. Descartes depended on the theory of emotions and thinking while presenting his ideas about consciousness. He developed the meditations in Meditation I to protect this phase of detachment from the senses. Meditation II, in the cogito, brings the realization of an initial reality. He also states that cogito's consequence is only understood because the mind perceives it "clearly and distinctly." Therefore, as the mark of reality, he creates distinct and straightforward logical perceptions, independent of the senses.

In Meditations III–VI, Descartes then unfolds the effects of simple and distinct perception, and in Principles I–II, he repeats and expands these outcomes. Finally, we consider these outcomes in Seconds.

In meditation one, he brings the senses into question, while in Meditation VI, he affirms that the senses are not supposed to be aware of the "essential essence" of external things. In this way, his role in the Meditations differs from that in the Laws. He permitted specific "simple natures" relating to corporeal objects to be regarded through the representations of the senses in that work.

In acquiring human intelligence, he attributed two functions to the senses.

He understood that the senses are essentially sufficient for the body to perceive advantages and harms. But, in reality, he found their ordinary task to be "to remind the mind of what is advantageous or detrimental to the composite of which the mind is a member," that is, to the composite of mind and body.

In reality, Descartes argued that all thoughts are, in some way, alive. The information would not come from the mind alone in Descartes' system of mental ability. The will belongs to that role. Unless the will has confirmed or rejected the substance provided by the intellect, a verdict, and therefore an example of (at least putative) knowledge, does not occur in this system. In reality, Descartes argued that all thoughts are, in some way, alive. Suppose perception (intellect, representation) is the essence of cognition.

There are many criticisms about this theory, we will conclude them in the following:

- It does not create a difference between thoughts and matters; instead, it makes this dichotomy the starting point in solving-knowledge theory issues. Thought and matter are ontological (things can exist), but consciousness describes our reactions that are being given in immediate, during sleepness, or while awake. It does not differ from animal to human. Human consciousness is the interaction between physiology and behavior; both physiology and behavior are the objective processes that can be observable (Fig. 5).
- Princess Elizabeth and Pierre Gassendi presented that if the soul affects the body, it must contact it. To do that, the soul must occupy space and have an extension. So, we conclude that the soul is physical by the criteria of Descartes's theory.

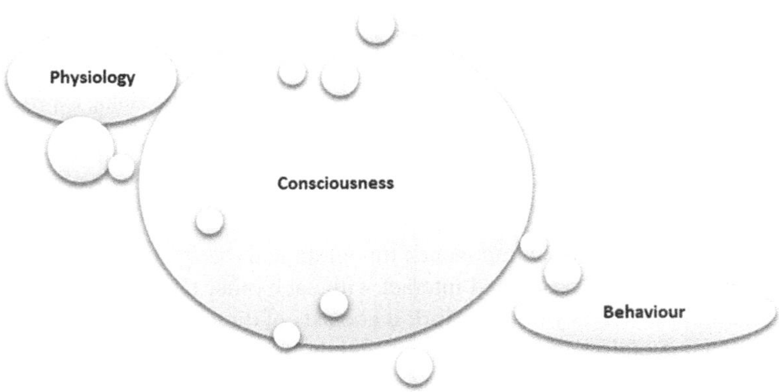

Fig. 5 Human consciousness is the result of the interaction between physiology and behavior

4 Spinoza Ideas

Spinoza sought to transcend Descartes' dualism. Rather than matter and mind becoming independent objects, Spinoza claimed that there is only one concept. Therefore, mind and extension are the properties of that one substance.

So, this merely changed the dualism of objects to a dualism of qualities. But unfortunately, it also retained Nature's mechanical concept presented by Descartes, rendering humankind exposed to an absolute mechanical fatalism.

5 Sigmund Freud on Consciousness

Sigmund Freud is considered the father of psychoanalytic theory, and we will depend on his ideas in classifying the levels of consciousness in the next chapter. He suggested a new and powerful way of thinking and investigating human thoughts, actions, and interactions (Hughes and Hoffman 2008).

He limited the role of the consciousness to that of an epistemological tool to know about certain areas of mental state, thus removing all ontological aspects. He thought that some states of mind could be found outside human awareness. That made him reshape the concept by rejecting the ontological principle of all states of consciousness.

According to Freud, "Consciousness is a fundamental and irreplaceable phenomenon regardless of the name" (Natsoulas 2001).

Awareness has been used to express what we mean when we are conscious of something and express the latent knowledge of something. "Conscious, if it is not burdened with any additional meaning, may indicate what is immediately subjectively and introspectively given an experience. So conscious can involve abstract ideas, obsessional preoccupation, and even hallucination" (Mcleod 2021).

> Humans are conscious of psychosis, dissociative states, intoxication, and so on. But each of these cases is in different mental organizations of experience, obeying different organization principles and existing with different levels of categorization and abstraction. The experience of consciousness may differ in each state, but consciousness as a subjective and introspective given is indivisible regardless of the state (Mcleod 2021).

> Freud describes consciousness as a quality and the capacity of transforming experienced activity into unconscious states. This is similar to how the different forms of energy are interchanging in physics. It could also play a role in inhibiting and restricting certain thoughts from becoming conscious. It also refers to transforming quantities of unconscious excitation into a qualitative experience of pleasure and unpleasantness (Mcleod 2021).

Freudian theory has two components, topographical theory and structural theory. These two theories overlap and interact with each other (Fig. 6).

Topographical Theory of the mind: It consists of three levels, and each of those three levels is specialized in some regions and thoughts. These three levels are:

1. Unconscious
2. Preconscious
3. Conscious

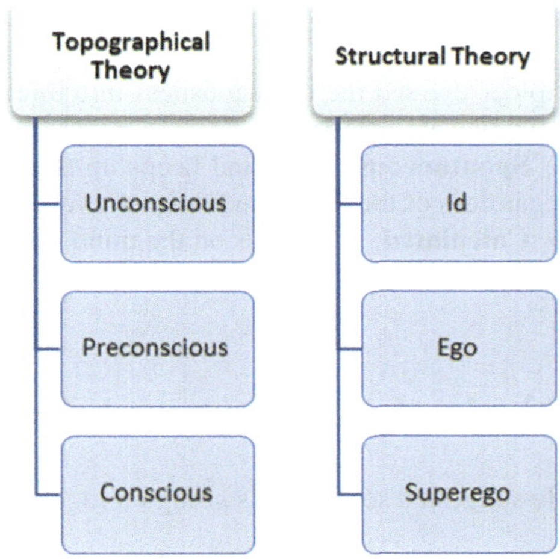

Fig. 6 The two components of the Freudian theory

We will discuss these levels in detail in chapter "Levels of Consciousness", which talks about the levels of consciousness as they are important and representative of human behavior. It is the most acceptable theory with some issues.

Structural Theory of the mind: These levels are also essential, and they overlap with the levels of topographical theory, so we will leave the details to chapter "Levels of Consciousness".

1. Id
2. Ego
3. Superego

The structural theory overlaps with the topographical theory to create the proper way of thinking and behaving and explain feelings and emotions.

6 Modern Theories of Consciousness

Even though the Freudian theory is still the most known explanation for consciousness, many schools and fields of psychology have started to develop new theories and levels to widen the science of consciousness. But all these theories are going in the same way, which is to understand the human self and answer the questions and queries of the consciousness. There are many models developed to explain the levels of consciousness depending on many criteria. We will discuss three models of these modern perspectives and the levels that they thought more suitable:

6.1 Holder's Levels of Consciousness

Holder divided the consciousness into three levels that give distinct differences in how they are reached.

Spontaneous: The mind keeps up the development and the progression of life regardless of the future and the past events.

Calculated: It depends on the mind's ideas of what is right and what is wrong.

Imposed: No awareness, and it ends with failure.

6.2 Levels of Consciousness, According to Barrett

He suggested seven levels going on logically (Fig. 7).

6.3 States of Consciousness According to Gibson

Dr. Bob Gibson stated four levels of consciousness: tiers of extrasensory awareness (Pons and Harris 2001; Tindall 1990). These levels interfere with the previous theories of the modern insight of consciousness. Gibson's insight focuses on some moments of a human's personal experience.

1. **Sleepness**: "Unaware of all surroundings, dreams may or may not occur."

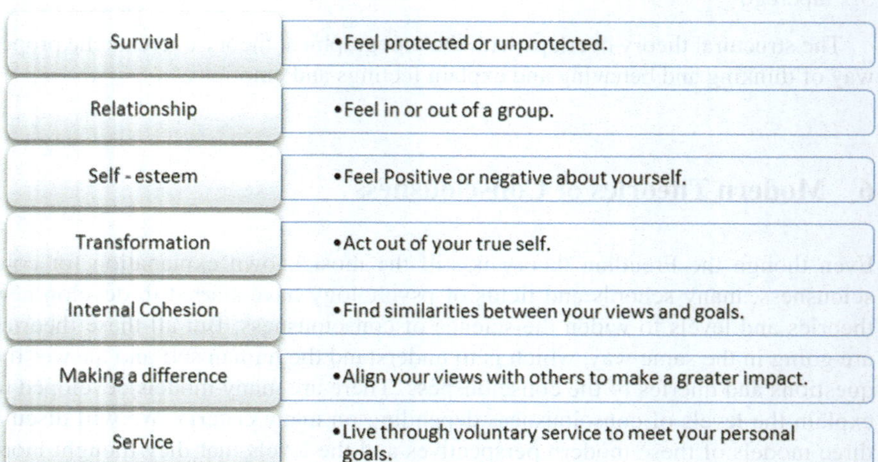

Fig. 7 Levels of consciousness according to Richard Barrett

2. **Waking sleepness**: "Normal tasks can be performed in sleep, but the individual is not receptive to what is taking place."
3. **The self-awareness**: "Able to identify surroundings and observe what is taking place."
4. **The objective awareness**: "Identify surrounding events without opinions or inputs."

Gibson's theory is more acceptable and accurate than the other theories in modern ideas. It is very descriptive and observable (Pons and Harris 2001; Tindall 1990).

6.4 8-Circuit Model of Leary About Consciousness

It is a combination of theories developed by Leary and Wilson. It consists of eight circuits, suggesting that the altered states of human consciousness come from eight brain regions.

This concept connects psychology to neurology and biology. It also incorporates elements of sociology, anthropology, physics, chemistry, and advanced mathematical formulas (Wilson 2008).

6.5 Morin's Integration

It is similar to the idea of Gibson, but it is concluded in four models as suggested in Fig. 8.

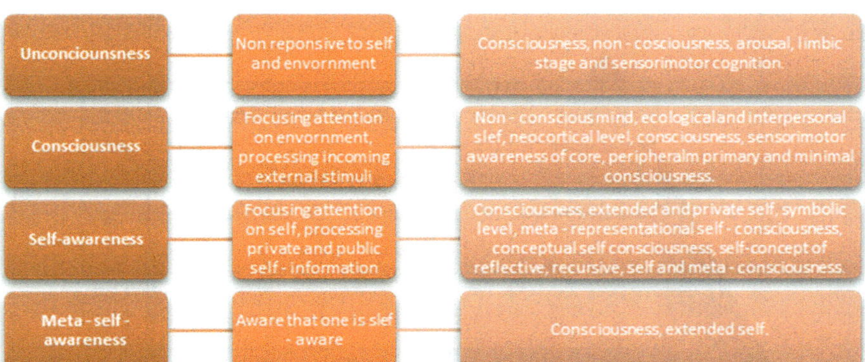

Fig. 8 Morin's hierarchy of conscious states

6.6 Moore and Gillette

It is an analytical description for imagination and the control of great power such as God and prophets (Lee 2012).

7 Developmental Psychology on Consciousness

A typical agreement remains on a few key points in the developmental growth of young children, such as the manifestation of the first social grin, the first voluntary moves, or the first sentences (Fig. 9).

A kid put in front of a mirror before the second year will usually whistle, coo, or show evident joy on seeing movements reflecting from the shiny surface of the mirror at them. However, the specular picture correlates dramatically different activities within 2 years.

The levels are described below in detail:

Level 0: **Confusion**: The person is characterized as unaware. The picture of speculation is confused with the reality of the world it portrays. It is seen as a simple continuation of, not a representation of, the universe.

Level 1: **Differentiation**: The person begins to be aware. This is the first degree of separation between the self and the world: a distinct self is articulated.

Level 2: **Situation**: The person can link the body image and the body's movements, so the person thinks systematically. There is no misunderstanding at this stage. The person knows that what is seen in the mirror is particular to himself.

Level 3: **Identification**: The person realizes the relationship between the self and the outside environment.

Level 4: **Permanence**: The person can keep thoughts by remembering and memorizing them over a long time.

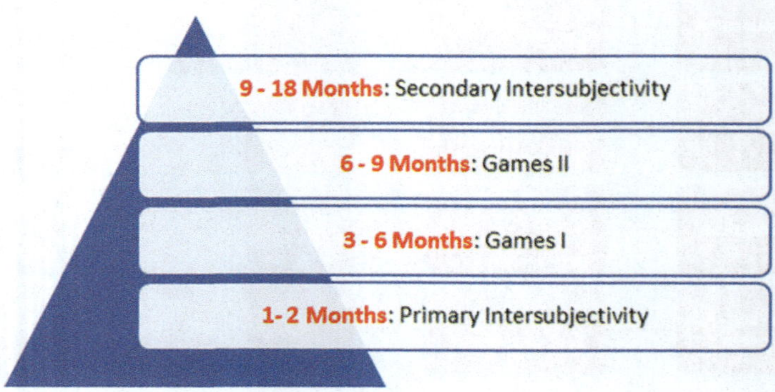

Fig. 9 Developmental psychology on consciousness

Level 5: **Self-consciousness or "meta" self-awareness**: At this level which is the last one, the person can realize the self in the other's eyes by being aware of him or herself.

Finally, the access and phenomenal consciousness factors depended mainly on the capacity of children to respond to their emotional states orally and to partake in pretend play as the primary evidence for children with the proper form of skills. The capacity to comprehend and use verbal representation and to pretend to play needs at least the production stage of different models. The way the universe (labeled as "real" or "fact") is portrayed by one model and another model is expected to display the orally defined universe or pretend scenario (labeled as "false" or "nonfactual"). In other terms, linguistic explanation and pretense both require clear factuality to be minimally transparent.

Piaget has a role in developmental psychology as well. He stresses that "meaningful inference" is the most common feature of the conscious systems, starting from the most superficial understandings of consciousness. It is defined by communicating concepts and putting them together in the form of a relation. Piaget argues in favor of causal isomorphism and participation in consciousness grabbing. The implication is a relation between interpretations. Suppose the causal coordination of behavior causes their material aims to be accomplished. In that case, the system of substantive consequences offers an aspect that is not known, either concerning the objectives or about the means used, in an acquisition that involves their importance, albeit a restricted acquisition: it is the determination of the motives, without which milestones reflect merely facts without importance.

8 Social Psychology on Consciousness

Psychology and especially social psychology are fundamental sociological sciences. In reality, social psychology has mainly grown through the efforts of sociologists even though it is confusing to be declared as a branch of sociology instead of psychology. For instance, Professor Ellwood prefers the designation "psychological sociology" for social psychology.

It is a valid interest to understand social activity and consciousness solely as a process of the individual's psychology to a particular portion of his environment, without thinking about the formation of the character of groups emerging from these reactions. The two sciences must remain distinct fields of study, notwithstanding the good offices and desires of sociologists (Bargh 2007).

The effect of one human upon another is often a matter of behavior. One individual activates, and the other reacts: We have the essence of social psychology in this process. Therefore, one organism activates another through some external indication or action; it is never perception. Stimulating and inhibiting both. At times, the responding actions may be followed by a social form of consciousness in the persons concerned; however, as far as we know, there is no immediate action of one person's consciousness on another's consciousness or actions.

In several quarters, an effort is underway to restrict the concept of culture and social science to forms of social contact where there is knowledge of others and social connections. This restriction is both non-essential and restricted from the point of view of the present job. As we just pointed out, consciousness has little impact, thus not clarifying something about human beings' reciprocal responses.

In social psychology, its function is analytical rather than explanatory, as in the non-social branches of research. No forces are binding the collective together, and no way of accessing the culture of thought or organized life, except in the most socialized and aware communities by the inner stimulation of one individual by the actions of another. In addition, if this term is intended to mean a form of stimulation other than physiological, it is not an "internal" inner stimulation, for there is no form of stimulation other than physiological. It would also be more fitting to accept that all types of animal life in which we see definite social activity are called in the field; that is, human reactions to each other.

In our present social psychology concept, social consciousness follows such social activity in the lower modes of existence, while of speculative importance, can be waived as non-essential. However, the social consciousness aspect in the subsequent chapters will not be overlooked by any means. On the contrary, it will be remembered wherever it is crucial in the entire situation or beneficial in determining the standards of conduct.

In our modern understanding of social psychology, the issue of whether social consciousness follows such social activity in the lower modes of existence, while of speculative concern, can be waived as non-essential.

At the time, these results were shocking (to social scientists and the general public) because they defined the strongly held belief of people that one's conduct was under the internal and conscious influence of one. Social psychology concentrates on interpersonal mechanisms of a comparatively higher order of complexity than cognitive psychology or cognitive neuroscience: for example, judgment, target pursuit over long periods, and actions in social interaction. Maybe as an inheritance or vestige of this long-standing conceptual status, therefore, by taking a central function for conscious (intentional, effortful, and conscious; see next section) decision and monitoring processes, social psychology appears to begin its study of any dynamic, significant phenomena. The study then has the effect of exploring the extent and function of the mechanism or phenomenon's non-conscious components.

In other typical fields of social psychological investigation, this phenomenon can also be observed. For example, early attribution theories were started by logical scientists with a model of humans, using effortful and deliberate "variance analysis" strategies to draw causality inferences. However, attribution theory moved to a more automated and less deliberative model as the research proof came in.

The motives and origins of their actions are mainly mysterious to individuals. In reality, new scientific research points to a profound and underlying dissociation between conscious consciousness and the mental mechanisms responsible for one's actions across many various fields of psychology; several of the wellsprings of actions seem to be invisible to conscious entry.

Knowledge of impact factors on decision and social behavior has long been an important research subject in social psychology. Starting with Nisbett and Wilson's 1977 seminal work, researchers found that individuals were ignorant primarily of actual strong effects on their decisions and actions. For example, laboratory subjects watched a work interview where the interviewee spilled coffee in one study. While others saw this tape without the event of the spill, attributes and stereotypes are first triggered in an off-hand, indirect way and then affect the subsequent decisions or actions of the individual, giving another indication of the dissociation between significant environmental factors and the perception of those factors by the person.

In several experiments, the key variable as to whether a person can regulate the external influence is not whether the person per se (i.e., whether it was subliminal or supraliminal) is aware of the controlling stimuli, but rather whether the person is aware of the possible effects of that stimuli.

Therefore, the subliminally presented priming stimuli have the same effect quality as those supraliminal presented, as long as the participant does not expect or appreciate that the stimulation could affect them.

Without the need for deliberate control of the person, action tendencies can be triggered and set into motion; even complex social activity will unfold without an act of will or knowledge of its origins. Proof from a broad spectrum of psychological investigation fields is associated with this proposition. For example, in cognitive science, clinical evidence from patients with frontal lobe lesions, action and goal-priming trials, cognitive neuroscience analyses on frontal lobe structure and function as well as distinct actional and semantic visual pathways, cognitive psychology experiments on working memory components, and the degree on conscious access to motor activity converge on the conclusion that complicated activity and other higher mental processes will proceed independently.

Indeed, neuropsychological research shows that such freedom is intended for the human brain. This is not to suggest, though, that knowledge does not exist or is simply an epiphenomenon. It also means that if any of these things can be done without deliberate choice or direction, then probably somewhere lies the object of consciousness (i.e., why it evolved).

While we do not yet know anything about the evolution of non-conscious target pursuit capacities, the most probable guess is that they evolve out of regular and reliable practice, much like other automated processes. In the case of automatic target pursuit, this suggests that the person most likely deliberately chooses to achieve that particular goal in that particular situation at one point, then picks it again, and so on until that representation of the target is so closely associated with that representation of the condition that the former is immediately associated with the latter.

The object of consciousness could be for the assemblage of complex unconscious abilities, which is why it evolved (Fig. 10). In accordance with the general plasticity of the growth of the human brain, human beings have the capacity to create even more complex automatic "demons" that sublimely suit their idiosyncratic

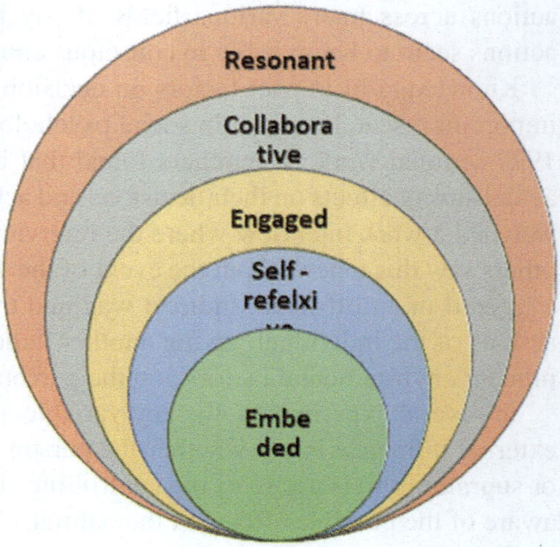

Fig. 10 The creation of human complex conscious abilities

environment, desires, and purposes, unlike even our closest primate family. Interestingly, then, one of the main aims of conscious processing at the person's level might be to remove the need for himself as far as possible in the future, freeing himself up for even greater things. Indeed, if the evolved aim of consciousness turned out to be the development of ever more complex non-aware systems, it would be ironic, considering the juxtaposition of automatic and aware systems in contemporary social psychology.

9 Neuropsychology on Consciousness

In neuropsychology, localization of the psychological roles in the brain has historically been focused on the finding that there is a deficiency or disorder of a certain psychological capacity in patients carrying a lesion in a specific cerebral area, whereas other abilities are retained. In the frontal lobe of the left hemisphere, the speech was located because lesions in that area induce expressive aphasia. However, comparable destruction in the right hemisphere does not have such an effect. In humans, cognitive perception relies on the function of the brain, so neuroscience can help to understand cognition (Gennaro 1996).

What would it feel like for cognition to be explained by neuroscience? In doing so, how much progress has neuroscience made? What challenges are they facing? How will those problems be met? What is the metaphysical value of its results? This entry answers these and similar questions.

Consciousness and unconsciousness are two separate stages of the entire body in regular daily life, based on various active modes of brain functioning, which alternate with the sleep-wake cycle somehow, but are partly independent of it. Compared to the pathological types of brain dysfunction that underlie coma unconsciousness,

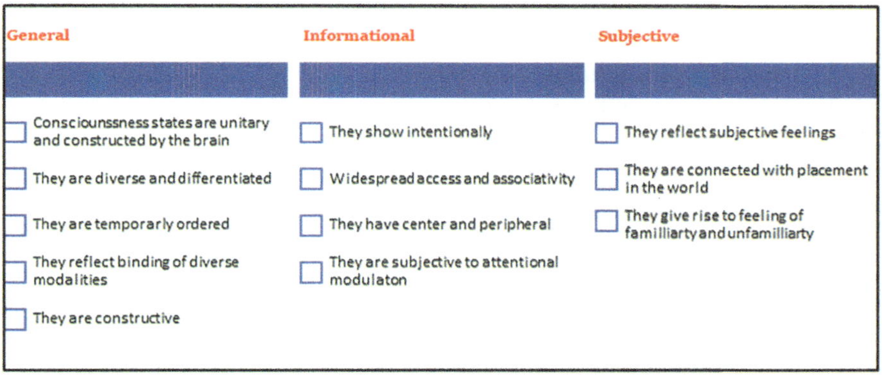

Fig. 11 The neuropsychology of consciousness; linking the brain to consciousness

the normal brain is still involved. The natural unconsciousness of dreamless sleep is a physiological state of brain functioning.

We will discuss the neuropsychology of consciousness with details in chapter "Brain and Mind", linking the brain to consciousness and discussing the differences between the mind and the brain (Fig. 11).

10 Hashim and Mustafa Theory of Postconscious

It is a newly developed theory in 2019. They defined consciousness as behavior and everything related to behavior is considered to be controlled by consciousness.

It is an extension of the Freudian theory, and it rearranges the areas of consciousness with an additional new concept called "Postconscious," which relates to future action and upcoming thoughts (Hashim and Ramadhan 2019; Hashim 2020).

This theory is novel, and studies have taken place to deeply discover this new level of consciousness. More details will be discussed in chapter "Levels of Consciousness" with the other levels of consciousness (Hashim and Ramadhan 2020).

Multiple Choice Questions

1. John Locke, in his theory, concluded that:
   (a) The consciousness is classified into three levels.
   (b) There is not any effect for the development of the person on his consciousness formation.
   (c) The topographic theory is more acceptable than others.
   (d) The actions and the ideas that the person makes during early life are essential for developing the future.

2. René Descartes concluded that the theory of the mind contains two domains that the consciousness resides in, which are:
   (a) Res Cogitans and Res extensa.
   (b) Preconscious and Postconsious.
   (c) Id and Ego.
   (d) Imagination and Reality.
3. The relationship between the mind and the body is described as:
   (a) A complementary relationship.
   (b) There is no relation between them.
   (c) How they can be related and how they affect one another. The characteristics of each one is different from the other.
   (d) Integrity relationship.
4. Topographical and structural theories have been described by:
   (a) Sigmund Freud
   (b) John Locke
   (c) Hashim and Mustafa
   (d) Spinoza
5. Holder divided the consciousness into three levels that give distinct differences in how they are reached. These three levels are:
   (a) Sleep, Walking, and Self-awareness.
   (b) Spontaneous, Calculated, and Imposed.
   (c) Id, Ego, and Superego.
   (d) Conscious, Postconscious, and Unconcious.
6. Psychoatomic (Quantum Non-Local) is:
   (a) Overmind, developed by consciousness maturity.
   (b) It is the adult personality and developed from first meeting experiences. It includes pleasure, reproduction, and nurture.
   (c) It is Buddha-Monad Mind, developed by consciousness maturation. It includes evolutionary consciousness, DNA–RNA brain feedbacks.
   (d) It is a Zen-Yoga mind–body connection developed by neurological somatic feedback and reprogramming. It includes the consciousness of the body.
7. Freud describes consciousness as:
   (a) A state of awareness of our external environment.
   (b) An imaginary concept to describe our awareness.
   (c) An origin for our reality.
   (d) A quality and the capacity of transforming experienced activity into unconscious states.
8. A psychiatry branch that describes the relationship between the brain and the psychology is called:
   (a) Neurology
   (b) Neurosciences
   (c) Neuropsychiatry
   (d) Neuropsychology

9. Postconscious is an extension of Freud theory and it has been described by:
   (a) Sigmund Freud
   (b) Spinoza
   (c) Hashim and Mustafa
   (d) René Descartes

Answers
1. (d)
2. (a)
3. (c)
4. (a)
5. (b)
6. (a)
7. (d)
8. (c)
9. (c)

References

Bargh JA. Social psychological approaches to consciousness. In: The Cambridge handbook of consciousness. Cambridge: Cambridge University Press; 2007. p. 555–69.
Butler J. Of personal identity. In: The analogy of religion. Berkeley: University of California Press; 1975. p. 100. Original reprinted in Perry, John, 1736, 'Personal Identity'.
Gennaro R. Consciousness and self-consciousness. Amsterdam: John Benjamins Publishing; 1996.
Hashim H. The role and the spectrum of the post-conscious in the control of behavior by geriatric subjects. NPG Neurologie - Psychiatrie - Gériatrie. 2020;20(119):297–302.
Hashim H, Ramadhan M. The need for developing a fourth level of awareness in human consciousness: unconsciousness, preconsciousness, consciousness and postconsciousness. J Psychol Psychother. 2019;9(3)
Hashim HT, Ramadhan MA. The concept of post-consciousness & its role in human behavior. J Conscious Explor Res. 2020;11:1.
Hughes H, Hoffman S. Consciousness and society. New Brunswick: Transaction Publication; 2008.
Kriegel U. Consciousness, theories of. Philos Compass. 2006;1(1):58–64.
Lee M. The sword: the development of consciousness. J Symbols Sandplay Ther. 2012;3(2):45–65.
Mcleod S. Unconscious mind | simply psychology. Simplypsychologyorg. 2021. <https://www.simplypsychology.org/unconscious-mind.html> Accessed 16 Dec 2021.
Miller G. What is the biological basis of consciousness? Science. 2005;309(5731):79.
Morin A. Levels of consciousness and self-awareness: a comparison and integration of various views. Conscious Cogn. 2006;15(2):358–71. https://doi.org/10.1016/j.concog.2005.09.006.
Natsoulas T. The freudian conscious. Conscious Emotion. 2001;2(1):1–28.
Nimbalkar N. John locke on personal identity. Mens Sana Monogr. 2011;9(1):268–75. https://doi.org/10.4103/0973-1229.77443.
Pons F, Harris P. Piaget's conception of the development of consciousness: an examination of two hypotheses. Hum Dev. 2001;44(4):220–7. https://doi.org/10.1159/000057061.
Reid T. Essays on the intellectual powers of man. Edinburg/London: John Bell/G.G.J. and J. Robinson; 1785. p. 317–8.

Tindall SC. Level of consciousness. In: Walker HK, Hall WD, Hurst JW, editors. Clinical methods: the history, physical, and laboratory examinations (PDF). Boston: Butterworths; 1990. p. 296–9. ISBN 0-409-90077-X.

Wilber K. Integral psychology: consciousness, spirit, psychology, therapy. Nova Religio. 2004;8(2):125–7.

Wilson RA. Cosmic trigger: final secret of the illuminati. Las Vegas: New Falcon Publications; 2008. p. 1–269. ISBN 978-1561840038

Brain and Mind

Hashim Talib Hashim and Adil Alhaideri

1 Brain Regions

The brain region is separated into left and right hemispheres. The corpus callosum joins these two hemispheres through white matter fibers (He et al. 2017).

The limbic system is responsible for memory and emotions (Fig. 1). In addition, it enables the hippocampus, the thalamus, the hypothalamus, and the brainstem to have complex connections. The limbic system does not have precise, specific anatomical limits but contains many substantial structures.

The amygdala is a series of nuclei found inside the brain (Fig. 2). It receives several sensory data modes as inputs. The outputs pass via the amygdala via the stria terminalis and ventral amygdalofugal pathway (Fig. 3).

Pathologies in the basal nuclei can be traced to certain movement conditions, the most notable being Parkinson's syndrome, which is due to defects in substantia nigra dopaminergic cells (Wardlaw et al. 2020; Mendez and Hong 1997; Rughani et al. 2015).

The neocortex is the human brain's most phylogenetically evolved organ. The intricate folding pattern causes a smaller cranial volume to be covered by an expanded cortical surface. The folding sequence that shapes the sulcal and gyral patterns remains strongly maintained in humans. This makes for a cortical anatomy nomenclature (Wardlaw et al. 2020; Herbet and Duffau 2020).

Just below the pons, the pyramids and pyramidal decussation are visualized ventrally. They are the descending corticospinal tracts (Jhawar et al. 2016).

The cerebellum has two hemispheres, connected via a midline structure called "the vermis." The cerebellar cortex has three membranes, in contrast to the neocortex of the cerebrum: molecular, Purkinje, and granular (Siegel et al. 1999). There

H. T. Hashim (✉) · A. Alhaideri
University of Baghdad, College of Medicine, Baghdad, Iraq
e-mail: adel.ferras1700b@comed.uobaghdad.edu.iq

H. T. Hashim, A. Alexiou (eds.), *The Psychology of Consciousness: Theory and Practice*, https://doi.org/10.1007/978-3-030-90692-4_2

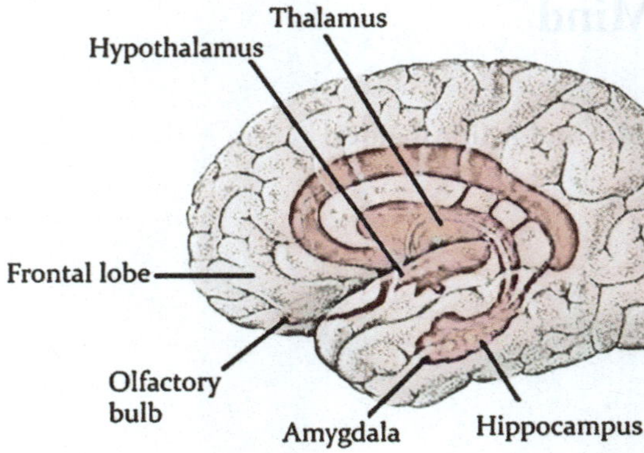

Fig. 1 The limbic system for the development of memory

Fig. 2 The amygdala

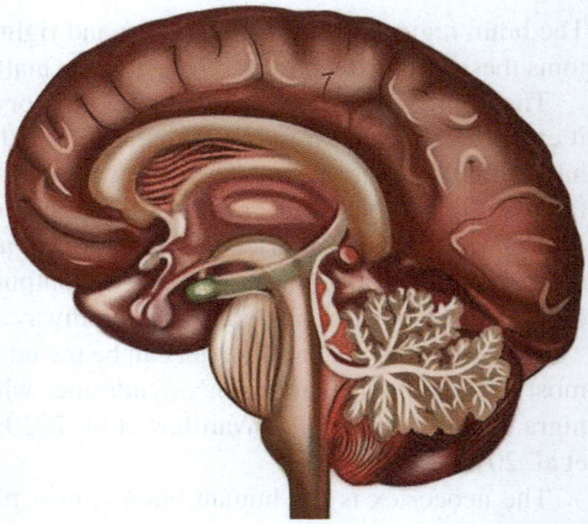

Fig. 3 The habenula in the epithalamus

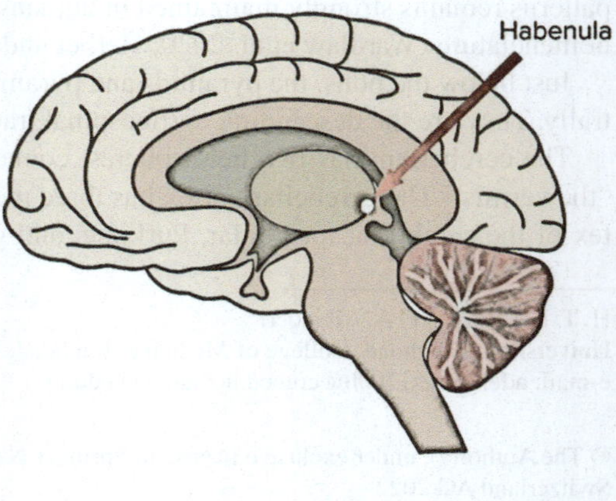

are afferent and efferent pathways inside the three cerebellar peduncles to and from the cerebellum. "The dissertation of neurologist Paul Broca, who researched the language disorders of stroke survivors, can be traced to some of the earliest contributions to modern language mapping" (Binkofski and Buccino 2004; Stein 2017).

The main sensory cortex, or Brodmann's areas 1–3, corresponds to the postcentral gyrus. That of the motor cortex corresponds to the homunculus obtained from awake mapping (Adrián-Ventura et al. 2019).

2 Brain–Mind Relationship

The word "mind" is widely used also within the scientific discourse to denote a wide spectrum of connotations. The mind can describe the human consciousness when the mind–brain connection is meant. However, consciousness, which can be seen as a confounding paradox, is itself impossible to describe, given that it constitutes perhaps the most immediate and intimately available feature of any person's existence (Kochiyama et al. 2018; Solomito et al. 2019).

While the brain is the body's most relevant organ in relation to the mind, more precise information is available (Duffau 2018; Solomito et al. 2019).

But if it is possible to clarify the mind–brain interaction, what sort of answer is being sought? It is important to consider previous scientific breakthroughs to address this key issue, culminating in a satisfactory explanation of interactions previously though difficult to describe (Forsell et al. 2020).

Although neuroanatomy provides the fundamental blueprint for nervous system information processing, the neural function is characterized by turbulent processes characteristic of complex systems. Thus, a complete comprehension of the brain as it applies to behavioral information would have to include appropriate accounts of synaptic dynamics and connectivity, as these elements are essential to conscious brain activity (Forsell et al. 2020).

It is possible to generalize this primary instance to all human contact. How much of the intended message is successfully conveyed on average among aware individuals communicating? We speculate that correspondence hovers between 50% and 85%, except in the most desirable situations of friends addressing their shared jobs or partners chatting about their family. If this is the case, 10% or less may be the normal human contact between two people (including casual interactions).

The immature brain has uncommitted, immature associations at birth, formed from within the primary. As a result, the senses of a newborn baby are overwhelmed with the outer world's many colors, noises, smells, tastes, and textures, as well as certain perceptions and emotions that originate from inside.

Increased activation contributes to increased synaptic complexity in the developing brain. As each person's perception is distinct, each brain–mind has different connectivity patterns. As Greenfield states, "The brain is made up of links between brain cells."

Implicit memory, which arises in the early years of childhood or out of intense dissociation produced through a traumatic experience, retains the habits that control our forms of being, behaving, and relating without conscious consciousness. It is Released in cases of misunderstanding which unpredictability, which inevitably provides the models in the consultation room for both the transition and counter-transfer encounter (Lei et al. 2017).

3 Consciousness in Medicine

It is more difficult to describe and classify the pathological state of consciousness, as demonstrated by the many words attributed by multiple observers to altered states of consciousness. For example, consciousness clouding, confusing conditions, delirium, obtundation, dementia, vegetative state, coma, hypersomnia, and brain death.

The meanings of these terms are different from one to another, and relating to a patient's state of consciousness, they can prove unreliable. However, it is desirable to describe some concepts as closely as possible (Lei et al. 2017).

There are a number of medical disorders and medicines that relate to the state of consciousness of an individual. Often it is reversible with compromised consciousness, while it is not at other times.

In medicine, we have two main levels for consciousness which are as follows (Huntley 2008):

- **Consciousness**: The patients are alert, awake, and responsive to stimuli like asking, talking, and expressing.
- **Unconsciousness**: The patient here is not alert and not awake. They are not responsive to any stimulus like sound or touch. Therefore, sleepiness is not unconscious.

Conclusion: there are disorders in psychiatry that can change and alter the consciousness and affect the brain, and at some times, it becomes a life-threatening issue.

The abnormal states of consciousness are as follows (Daroff et al. 2012; Cecil et al. 2012; Rakel and Rakel 2011):

- **Confusion**: Being confused means being disoriented to time, place, and person.
- **Lethargy** (tiredness): The patient is tired and likes to fall asleep.
- **Delirium**: Acute confessional condition, marked by diminished cognition, in a specific concentration. Alcohol withdrawal, recreational drugs, medicines, cancer, organ dysfunction, and severe infections can cause.
- **Obtundation**: This means sluggish responses to stimuli and a decline in alertness.
- **Coma**: The coma means reduced brain activity and lack of response from the patients to any stimuli, even some reflexes.

There are two popular and useful tools used in assessing the level of consciousness for any patient which are (Opara et al. 2014; Reith et al. 2017):

- ***Grady Coma Scale (GCS)***: Starts from 1 to 5 depending on the level of consciousness and the patient's responses to the stimuli. Grade 1 is active and Grade 5 means that the patient is in a coma.
- ***Glasgow Coma Scale (GCS)***: Starts from 1 to 15 depending on specific criteria and stimuli and measures the responses. Less than 3 means that the patient is in a coma or brain death, while 9 and above means that the patient's neurological condition is good.

Multiple Choice Questions

1. .... is involved in memory and emotions.
   (a) Cerebellum
   (b) Limbic system
   (c) Thalamus
   (d) Hypothalamus

2. Although neuroanatomy provides the fundamental blueprint for nervous system information processing, the neural function itself is characterized by:
   (a) Turbulent processes characteristic of complex systems.
   (b) Mind–brain relationships.
   (c) Conscious reactions.
   (d) Moving the body parts.

3. The differences between human and other animal brains are:
   (a) The neocortex is the human brain's most phylogenetically evolved organ.
   (b) The animals lack cerebellum.
   (c) Humans have bigger brains than animals.
   (d) Humans have more complicated brains than animals.

4. There are two hemispheres in the cerebellum, connected by a midline structure called:
   (a) The limbic system
   (b) The vermis
   (c) The sulci
   (d) The gyrus

5. The patient is alert and responsive:
   (a) Unconscious
   (b) Coma
   (c) Conscious
   (d) Sleepy

6. The patient is not responsive at all and may have reduced brain activity:
   (a) Coma
   (b) Lethargy
   (c) Unconscious
   (d) Confusion

7. Delirium is:
   (a) Extreme drowsiness, listlessness, and apathy followed by decreased alertness are identified.
   (b) Acute confessional condition, marked by diminished cognition, in specific concentration, sleep-wake period modification, hyperactivity (agitation) or hypo-activity (apathy), cognitive abnormalities such as hallucinations (seeing objects that are not there) or illusions (false beliefs), as well as pulse rate and blood pressure dysfunction.
   (c) No response at all.
   (d) Insensitive.

8. Obtundation is:
   (a) This is a condition of unresponsiveness, including stimulus.
   (b) Awakeness and alertness.
   (c) With sluggish responses to stimuli, a decline in alertness needs repeated stimulation to sustain concentration.
   (d) Acute confessional condition, marked by diminished cognition, in specific concentration, sleep-wake period modification, hyperactivity (agitation) or hypo-activity (apathy), cognitive abnormalities such as hallucinations (seeing objects that are not there) or illusions (false beliefs), as well as pulse rate and blood pressure dysfunction.

9. GCS of 14 is considered as:
   (a) Good
   (b) Fair
   (c) Poor
   (d) Excellent

10. In neuropsychology, the localization of psychological roles in the brain has historically been focused on the finding that there is:
   (a) A deficiency or disorder of a certain psychological capacity in patients carrying a lesion in a certain cerebral area, whereas other abilities are retained.
   (b) A balance between the chemical and physical components of the brain.
   (c) A reaction between the neurotransmitters and the body's status.
   (d) A balance between the sympathetic and parasympathetic systems.

Answers

1. (b)
2. (a)
3. (a)
4. (b)
5. (c)
6. (a)
7. (b)
8. (c)
9. (a)
10. (a)

References

Adrián-Ventura J, et al. Linking personality and brain anatomy: a structural MRI approach to reinforcement sensitivity theory. Soc Cogn Affect Neurosci. 2019;14(3):329–38.

Binkofski F, Buccino G. Motor functions of the Broca's region. Brain Lang. 2004;89(2):362–9. https://doi.org/10.1016/S0093-934X(03)00358-4.

Cecil RL, Goldman L, Schafer AI. Goldman's Cecil medicine. 24th ed. Philadelphia: Elsevier/Saunders; 2012.

Daroff R, Fenichel G, Jankovic J, Mazziotta J. Bradley's neurology in clinical practice. 6th ed. Philadelphia: Elsevier/Saunders; 2012.

Duffau H. The error of Broca: from the traditional localizationist concept to a connectomal anatomy of human brain. J Chem Neuroanat. 2018;89:73–81.

Forsell L, et al. BrainWiki—a wiki-style, user driven, comparative brain anatomy tool. Front Neuroanat. 2020;14:548172.

He Q, Turel O, Bechara A. Brain anatomy alterations associated with social networking site (SNS) addiction. Sci Rep. 2017;7(1):1–8.

Herbet G, Duffau H. Revisiting the functional anatomy of the human brain: toward a meta-networking theory of cerebral functions. Physiol Rev. 2020;100(3):1181–228.

Huntley A. Documenting level of consciousness. Nursing. 2008;38(8):63–4. https://doi.org/10.1097/01.NURSE.0000327505.69608.35.

Jhawar SS, Nunez M, Pacca P, Voscoboinik DS, Truong H. Craniovertebral junction 360°: a combined microscopic and endoscopic anatomical study. J Craniovertebral Junction Spine. 2016;7(4):204–16. https://doi.org/10.4103/0974-8237.193270. PMID: 27891029; PMCID: PMC5111321.

Kochiyama T, et al. Reconstructing the neanderthal brain using computational anatomy. Sci Rep. 2018;8(1):1–9.

Lei Y, et al. The brain interstitial system: anatomy, modeling, in vivo measurement, and applications. Prog Neurobiol. 2017;157:230–46.

Mendez I, Hong M. Reconstruction of the striato-nigro-striatal circuitry by simultaneous double dopaminergic grafts: a tracer study using fluorogold and horseradish peroxidase. Brain Res. 1997;778(1):194–205. https://doi.org/10.1016/s0006-8993(97)01055-x.

Opara JA, Małecka E, Szczygiel J. Clinimetric measurement in traumatic brain injuries. J Med Life. 2014;7(2):124–7.

Rakel RE, Rakel D. Textbook of family medicine. 8th ed. Philadelphia: Elsevier/Saunders; 2011.

Reith FCM, Lingsma HF, Gabbe BJ, Lecky FE, Roberts I, Maas AIR. Differential effects of the Glasgow coma scale score and its components: an analysis of 54,069 patients with traumatic brain injury. Injury. 2017;48(9):1932–43. https://doi.org/10.1016/j.injury.2017.05.038.

Rughani AI, Penar PJ, Gest TR. Brain anatomy. Medscape; 2015.

Siegel GJ, Agranoff BW, Albers RW, Fisher SK, Uhler MD. Basic neurochemistry, molecular, cellular, and medical aspects. 6th ed. Philadelphia: Lippincott, Williams & Wilkins; 1999.

Solomito MJ, Reuman H, Wang DH. Sex differences in concussion: a review of brain anatomy, function, and biomechanical response to impact. Brain Inj. 2019;33(2):105–10.

Stein J. Sensorimotor control. Reference module in neuroscience and biobehavioral psychology. Amsterdam: Elsevier; 2017. https://doi.org/10.1016/B978-0-12-809324-5.06855-3. ISBN 9780128093245.

Wardlaw JM, et al. Perivascular spaces in the brain: anatomy, physiology and pathology. Nat Rev Neurol. 2020;16(3):137–53.

Levels of Consciousness

Hashim Talib Hashim and Mustafa Ahmed Ramadhan

1 Introduction

Before we start with the levels, we should explain Id, Ego, and Superego according to Freud's theory (Fig. 1):

- **Id (Meeting basic needs):** This is the most primitive part of the personality; it contains the most animalistic urges. For example, eating, drinking, and having sex. It provides gratification for human needs. If humans do not get their needs, they will go into a disorder like anxiety, tension, and anger. This level is presented at birth. The examples for that are: when you are hungry, you will enter the kitchen and search for food, open the food pots, search in the fridge for anything that can be eaten, when you do not find anything to keep you settled, the situation will be undesired. It is the same situation when your hungry baby is crying; he will not stop until he gets his needed food.
- **Ego (Dealing with reality):** This concept tries to meet the desire that Id present in a socially acceptable way. This can mean delaying the gratification and helping to get rid of disordered thoughts like tension or anger. This level is developed at birth. For example, when someone is Muslim, during Ramadhan, he fasts during the day and should wait for the sunset to eat and drink. So, the Id desires to eat and drink, and the Ego is to wait until sunset.
- **Superego (Adding Morals):** It is lastly developed; it depends on morals and judgments about right and wrong. The Id and Ego can reach the same point as the Superego does, but Superego depends on morals while the Ego depends on the thinking and the consequences of the actions. So, losing the Superego makes Ego dominate, and losing Ego makes Id dominate. It begins to develop at about 6 years of age. For example, when you want to steal from your family and know

H. T. Hashim (✉) · M. A. Ramadhan
University of Baghdad, College of Medicine, Baghdad, Iraq

H. T. Hashim, A. Alexiou (eds.), *The Psychology of Consciousness: Theory and Practice*, https://doi.org/10.1007/978-3-030-90692-4_3

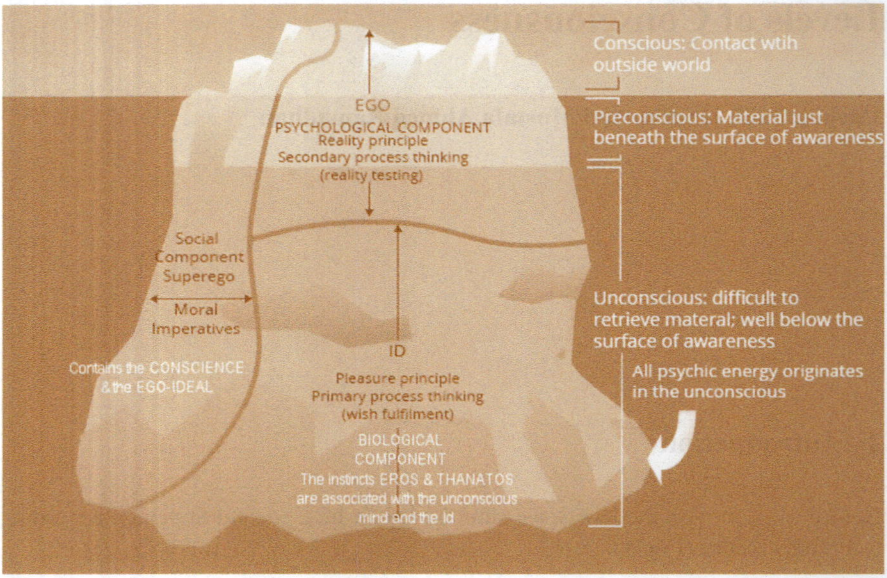

Fig. 1 Freud divides the mind into three levels according to his topographical theory: unconscious, preconscious, and conscious. These three levels interfere with structural theory components: Id, Ego, and Superego

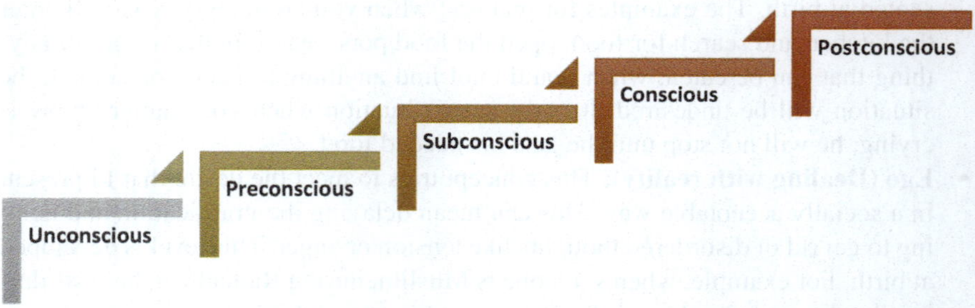

Fig. 2 The levels of consciousness

that no one will note or judge you, you don't do this because you know that stealing is wrong.

We will start in a sequence from the first level: unconscious, preconscious, subconscious, conscious to postconscious (Fig. 2).

2 Preconscious

It refers to the thoughts that humans are not actively thinking of but can be retrieved easily once the human wants that; it contrasts with the unconscious, whose thoughts are repressed and cannot be retrieved or remembered easily.

Preconscious was developed in the psychoanalytic theory by Freud. However, eclectic therapists and others may use the concept in their ideas.

The thoughts at this level are not in our current thinking, but humans can think of them quickly by triggering them using a stimulus of something that links them. These thoughts are pulled out of the unconscious to be stored in the preconscious and then transformed into conscious when needed. Ego and Superego control it only without Id (Fazekas and Overgaard 2016).

A preconscious mind is a convenient thing to have. It provides several valuable things that the conscious mind cannot handle alone.

The type of memories stored here is the long-term memories that align with the unconscious. They are both unconscious, and preconscious includes the memories that can be retrieved longer than a few minutes. The repressed memories can be recalled without a potent stimulus or trigger. So, preconscious is the caller one that connects unconscious with conscious.

Preconscious provides solutions for worries and problems without being aware of them. The process is going on at this level until a solution comes up. Like it works behind the scenes (Fazekas and Overgaard 2016).

The preconscious also contains intuitive thoughts. For example, when you meet someone, your mind will judge it as good or not good. This process occurs in the preconscious, depending on your brain's criteria to judge people and other thoughts and situations (Roulin and Ramelet 2014).

Adults have more active preconscious infants and teenagers because it depends on experience and knowledge taken throughout life and ages. These ideas support self-protection and achievements.

Therapies in psychiatry and psychology target preconscious because they need to change bad thoughts to overcome good ones. The therapists can usually prompt the patient to remember facts and events that will help him build his own conclusions. If the psychologists already know your history, they may have some ideas about what kind of experience he might suffer to explain the current situations (Overgaard and Overgaard 2010).

There is a type of therapy called "Connecting memories." In this therapy, the therapist tries to connect the patient's memory with another one to use the same solution for the previous problem in the current one. It is a connection between the preconscious and conscious. The ability to solve problems and issues in the conscious mind depends on using the information and thoughts in the preconscious. If humans cannot, they will be without benefit.

In conclusion, preconscious is characterized by reality testing, recallable memories, and links to word presentation—the key distinction from unconscious contents.

Words presentations are memory traces that were at one-time perception and therefore can become conscious again.

According to the French psychosomatic school, sound organization of preconscious is aligned with sound mentalization, whereas occasional long-term weakness can lead to “Operative Functioning” according to the French psychosomatic school. Inversely, the sound functioning of the preconscious guarantees a rich capacity for fantasizing (Overgaard and Overgaard 2010).

3 Subconscious

It is part of the mind that is not currently in focal awareness. But it is the powerful secondary system that runs everything in a human’s life. The subconscious is a bank of data for everything that is not in the conscious; it stores beliefs, thoughts, memories, and previous experiences. So, everything that humans see, think, and do is stored here (Fig. 3).

Subconscious is a guidance system that monitors information and actions coming out of senses and opportunities. Thus, the relationship between the subconscious and conscious is bidirectional (Nelson 2005).

Emotions control communication between the subconscious and conscious, so moving thoughts to another is done under their control. Only thoughts transferred with guanine emotions make it possible to be back for the mind. And only the emotions that are backed up by solid emotions are stored in this level. Both negative and positive emotions have a role in this level, and the negative ones are dominant (Nelson 2005).

Fear and negative talking to self are contributing factors in dominating the negative emotions, so to be in a good situation, you have to eliminate these thoughts. Because sometimes, fears and bad thoughts can be real and maybe come true (Hawkins 2015).

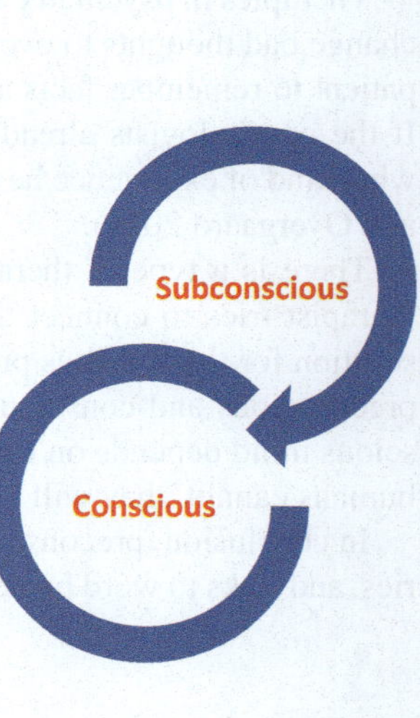

Fig. 3 The subconscious includes everything that is not in the conscious

There is a solution that can be used to get rid of bad thoughts through countering techniques. By confronting these ideas with extreme positive counter thoughts. For example, an employee on his first interview with a human resources manager in a company, when he thinks that he will not be competitive and efficient for the position, immediately face these thoughts with good ones by thinking that you will convince the manager by his skills. There is another technique called "Delete Button." When humans think of bad thoughts, they immediately delete them and replace them with good ones (Hawkins 2015).

Desire has a role in this level; depending on the desire, the subconscious tries to do anything to reach the desire that the mind seeks. For example, a very brave commander in a war is facing a weak army. The weak army makes noise and hits the drums to make the commander think they are not afraid of him; in this battle, the commander lost because of his thoughts.

The subconscious mind combines what one sees, experiences, and any knowledge that the mind absorbs that it does not otherwise actively interpret to make coherent sense. The conscious mind cannot necessarily process disconnected knowledge since it will be abundant information. Still, the subconscious mind retains this data where the conscious mind can recover it as it has to protect itself for survival (and for other purposes, such as solving puzzles).

The subconscious mind retains knowledge that the conscious mind does not automatically process with complete comprehension, but when "reminded" by the conscious mind, or by an astute psychoanalyst who can pull out the information contained in the subconscious, taking it to the conscious consciousness of the person, it retains the information for later retrieval.

The subconscious mind retains knowledge that the conscious mind does not automatically process with complete comprehension, but when "reminded" by the conscious mind, or by an astute psychoanalyst who can pull out the information contained in the subconscious, taking it to the conscious consciousness of the person, it retains the information for later retrieval.

De Becker tapped into the victim's mind about her "prior knowledge of the subconscious mind that prompted her to act unconsciously," causing her to know that the attacker would kill her. The analyst took her conscious mind to consider by eliciting her initial "inner thoughts/voice" HOW her subconscious acted on her conscious mind. Via a sequence of events that eventually drove her subconscious mind to function in such a way as to prevent her from being killed. Gavin was able to evoke awareness from her subconscious mind of a threatening condition that prompted her conscious mind to intervene to save her by her simple instinct of survival, taking to the conscious mind of the survivor that it was the "subtle warning that alerted her." The survivor characterized this as an unrecognized terror that drove her to act, still consciously uncertain of exactly WHY she was afraid. Her conscious mind had said the words, "I swear I would not harm you, although her subconscious mind measured the circumstance even more easily than the conscious mind could make sense of WHY there was terror. The victim said that" the beast took over inside her. It is impossible to separate the unconscious from the subconscious. Some authors have actually noted that they are used interchangeably in common parlance and by many experienced writers. It's good to think about conscious consciousness as the tip of an iceberg, as in discriminating between suppressed and suppressed: it's so apparent over the sea. Although the unconscious and subconscious are far greater than what the eye can see together, both remain below what is readily apparent. Thus, knowing their relative inaccessibility is the only practical way they can be set apart (Wu et al. 2019).

In brief, you will possibly recognize from where your thought, instinct, or inspiration is subconsciously derived through some introspection. Yet, it would be much more difficult to identify the source of present-day activity that simply does not make much sense to you for what is unconscious to you, the bottom-most portion of the iceberg. In addition, though, it is much more likely that you can easily unveil its roots with the help of a mental health provider (Wu et al. 2019).

Many ways to manipulate the subconscious mind specifically, including the following:

- Affirmations
- Autosuggestion
- Binaural beats
- Hypnosis
- Placebo
- Priming
- Subliminal perception
- Subliminal stimulation
- Suggestion

4 Conscious

The conscious mind acts as the current situation for the human. It is simply our current behaviors and actions. These two functions can be consciously responsible for in general:

- Its ability to direct human focus.
- Its ability to create images that are not real.

The conscious mind acts as a scanner for the human self. It perceives events, triggers the need to react, and then depends on the importance of that event.

It can store these events either in the Unconscious or Subconscious area of the mind where they will be available to human needs, and humans can recall them once they want (Young and Rund 2010).

The conscious mind includes everything inside of human awareness in the psychoanalytic theory. So, it makes the mind think and act rationally by mental processing.

The conscious mind controls sensations, perceptions, memories, feelings, and fantasies inside a human's current awareness. The closer level of awareness to conscious is Preconscious or Subconscious, and it includes things that we are not aware of in the moment of thinking but can be easily recalled in the conscious mind (Young and Rund 2010).

The conscious mind is the objective and thinking mind of humans. It has no memory, and it can only show one thought at one time. This level has four essential functions in particular, which are considered under the two general functions above:

- Identification of incoming information. This information can come from the five senses: sight, sound, smell, touch, taste, and feeling. So, it is observing and classifying what is happening in a continuous pattern around the human.
- Comparison between the current situation with similar situations in subconscious or preconscious. For example, when you see the face of someone, you will immediately compare this face with previously known faces to see if you have met them before; that is why you sometimes see a face and think it is familiar to you. Another example is when you work on a machine and are afraid of hurting your hand, your subconscious bank will tell you that it is safe because you wear protective gloves.
- Analysis. The conscious mind analyzes the data received by the senses and either accepts it or rejects it. It can deal with one action in a moment, like either positive or negative, yes or no, good or bad. So, when your senses send a message to anything the conscious is dealing with to the mind. It analyzed these data depending on its criteria and fundamentals to process to the next function. For example, you are walking in the street, and suddenly, you hear a high voice coming out of a house in that street, this voice will be entered in an analysis by the conscious.
- Deciding. It is related to the previous function of data analysis; in this function, conscious decide what is good and bad, negative, and positive. For example, the same situation above, after analyzing the data of that voice, your mind will decide that this voice is an expression of either sadness or happiness, depending on the criteria of the voice and links it to the bank in the subconscious (Zelazo et al. 2007).

5 Postconscious

It is an imaginary (Unreal) level of consciousness that's created to explain the actions that the other levels of consciousness cannot explain. It deals with future events and thinking. It is not a defense mechanism, but it solves a state of being unsure or confused about two things: right and wrong. It makes the mind comfortable to avoid unwanted psychological collapses. For example, a person is about to do something wrong or is confused between two choices. This person would struggle between the conscience that allows him to go on and the unconsciousness that stops him from doing so anytime you want to do something wrong. The postconsciousness that offers the reason to do the wrong things will overcome this dispute, and it is essential and will never be repeated. As humans, we have certain situations and attitudes that other forms of consciousness cannot be clarified since the four levels of perception in our consciousness influence all our behavior and reactions. All of these responses are that our attitudes and disposals alter whether we decide to make a mistake or it has just been done. As a consequence of our behavior, which is the failures, the explanation of these changes in our reactions is regulated as we try to approve the fourth stage of consciousness, the postconsciousness, in this analysis. As a consequence of our behavior, which is the failures, the explanation of these changes in our reactions is regulated as we try to approve the fourth stage of consciousness, the postconsciousness in this analysis (Hashim and Ramadhan 2019; Hashim and Ramadhan 2020; Hashim 2020).

The postconscious advises you to do these reactions or instruct the body to demonstrate these reactions (in a detailed description) to overlap the doubts and worries that might exist or reassure you that you will not be discovered or captured. So, it wants to defend you, and it tries to get you as happy as possible at the same time. In this phase, age and gender do not influence because it happens to anyone at any age, regardless of gender. Under the same names in Freud's principles of consciousness, we should not accept these emotions because they are distinct.

The postconscious is a small state that in elderly people may have less value and can cause other effects that vary from their usual roles in their brains. Over the life of an older person, certain thoughts and perceptions may change: some of them will vanish, and new ideas will arise, describing odd behaviors at this point. The figures below compare the four levels of consciousness in the factors that almost control them and the differences between them (Figs. 4 and 5).

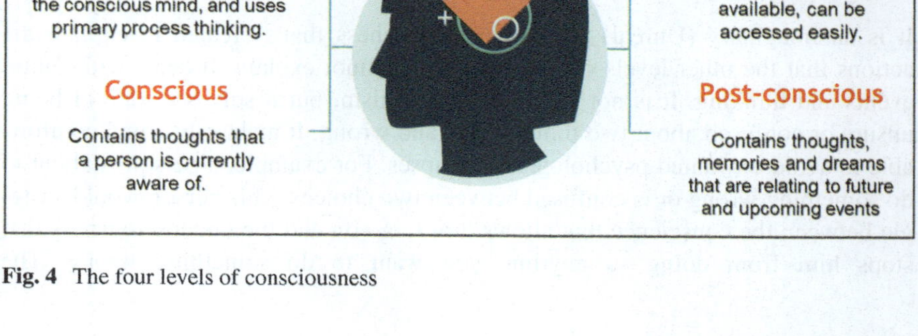

Fig. 4 The four levels of consciousness

COMPARISON

	Unconscious	Preconscious	Conscious	Postconscious
Influenced by external reality	✗	✓	✓	✓
Presents at birth	✓	✗	✗	✗
Associated with moral values and conscience	✓	✓	✓	✗
Id involvement	✓	✗	✗	✗
Ego involvement	✓	✓	✓	✓
Superego involvement	✓	✓	✓	✓

Fig. 5 Comparison of the four levels of consciousness

Multiple Choice Questions

1. It is the most primitive part of the personality; it contains the most animalistic urges:
   (a) Ego
   (b) Superego
   (c) Id
   (d) Subconscious

2. When you want to steal from your family and know that no one will note or judge you, you don't do this because you know that stealing is wrong. This presents:
   (a) Id
   (b) Superego
   (c) Ego
   (d) Consciousness

3. It refers to the thoughts that humans are not actively thinking of but can be retrieved easily once the human wants that; it contrasts with the unconscious, whose thoughts are repressed and cannot be retrieved or remembered easily.
   (a) Preconscious
   (b) Consciousness
   (c) Postconscious
   (d) Unconscious

4. The relationship between the subconscious and conscious is:
   (a) Complementary
   (b) Opposite
   (c) Integrated
   (d) Bidirectional

5. ………….. acts as a scanner for the human self. It perceives events, triggers the need to react, and then depends on the importance of that event.
   (a) Conscious
   (b) Subconscious
   (c) Preconscious
   (d) Postconscious

6. ……….. is an imaginary (unreal) state of consciousness.
   (a) Subconscious
   (b) Conscious
   (c) Preconscious
   (d) Postconscious

7. Subliminal stimulation is one of the ways that manipulate:
   (a) Conscious
   (b) Subconscious
   (c) Preconscious
   (d) Postconscious

8. One of the following is present at birth:
   (a) Preconscious
   (b) Conscious
   (c) Unconscious
   (d) Postconscious

9. All the following has an Id involvement except:
   (a) Preconscious
   (b) Conscious
   (c) Unconscious
   (d) Postconscious

10. Postconscious is:
   (a) An area between conscious and unconscious.
   (b) A complimentary area for consciousness.
   (c) A part of the unconscious.
   (d) An integrated Level of consciousness

Answers

1. (c)
2. (b)
3. (a)
4. (d)
5. (a)
6. (d)
7. (b)
8. (c)
9. (c)
10. (a)

References

Fazekas P, Overgaard M. Multidimensional models of degrees and levels of consciousness. Trends Cogn Sci. 2016;20(10):715–6.
Hashim H. The role and the spectrum of the post-conscious in the control of behavior by geriatric subjects. NPG Neurologie - Psychiatrie - Gériatrie. 2020;20(119):297–302.
Hashim H, Ramadhan M. The need for developing a fourth level of awareness in human consciousness: unconsciousness, preconsciousness, consciousness and postconsciousness. J Psychol Psychother 2019;9(3).
Hashim HT, Ramadhan MA. The concept of post-consciousness & its role in human behavior. J Conscious Explor Res. 2020;11:1.
Hawkins DR. Transcending the levels of consciousness: the stairway to enlightenment. Carlsbad: Hay House; 2015.
Nelson K. Emerging levels of consciousness in early human development. In: The missing link in cognition: origins of self-reflective consciousness. Oxford: Oxford University Press; 2005. p. 116–41.
Overgaard M, Overgaard R. Neural correlates of contents and levels of consciousness. Front Psychol. 2010;1:164.
Roulin M-J, Ramelet A-S. Behavioral changes in brain-injured critical care adults with different levels of consciousness during nociceptive stimulation: an observational study. Intensive Care Med. 2014;40(8):1115–23.
Wu G-R, et al. Modulation of the spontaneous hemodynamic response function across levels of consciousness. NeuroImage. 2019;200:450–9.
Young JL, Rund D. Psychiatric considerations in patients with decreased levels of consciousness. Emerg Med Clin North Am. 2010;28(3):595–609.
Zelazo PD, Gao HH, Todd R. The development of consciousness. In: The Cambridge handbook of consciousness. Cambridge: Cambridge University Press; 2007. p. 405–32.

Sleep and Dreaming

Saleh Abdulkareem Saleh

1 Sleep

The quality of sleep is essential for survival as food and water. Most body systems enter the anabolic state during sleep and help maintain normal body function (Sollars and Pickard 2015).

1.1 Structures Involved in Sleep

Many parts of the brain control sleep, like the hypothalamus.

The brainstem, thalamus, and pineal body that creates melatonin are also crucial for controlling sleep in both directions, REM and NREM (Roenneberg et al. 2013).

1.2 Circadian Rhythms and Sleep

The term circadian means "around the day" in Latin, so circadian rhythms are physical, mental, and behavioral changes that follow a daily cycle by responding to light and darkness. They are also present in plants, animals, and microbes. At the same time, biological clocks are a collection of specific proteins that interact with cells throughout the body and act as innate timing devices to produce and regulate circadian rhythms (Schupp and Hanning 2003).

The circadian rhythm controls sleep patterns. SCN is triggered by incoming light from the optic nerve and controls the production of the melatonin

S. A. Saleh (✉)
University of Baghdad, College of Medicine, Baghdad, Iraq

H. T. Hashim, A. Alexiou (eds.), *The Psychology of Consciousness: Theory and Practice*, https://doi.org/10.1007/978-3-030-90692-4_4

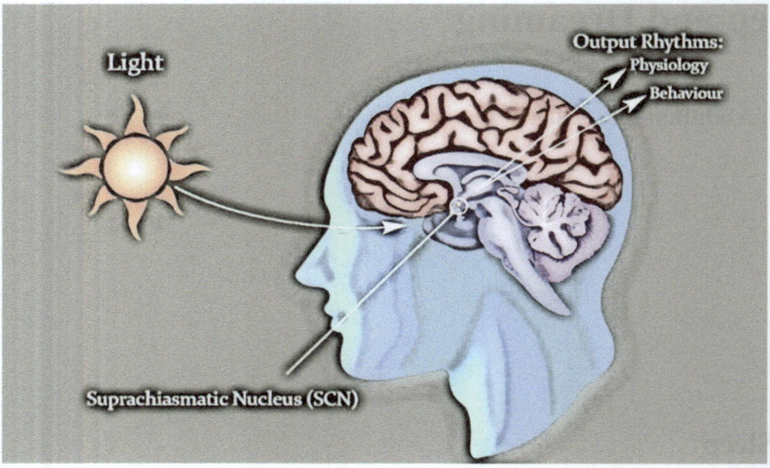

Fig. 1 The effect of light on SCN

hormone—responsible for sleeping. When the incoming light decreases the SCN, it stimulates the brain to produce more melatonin to make us drowsy and prepared for sleep (Fig. 1). In this way, the circadian rhythm controls sleep and wakefulness throughout the day and night to create a stable regular cycle. Disruption to this circadian system can result in sleep difficulties and disturbances.

1.3 Sleep Types

Two types of stages for sleep are discussed. They are (Fig. 2):

1. Non-Rapid Eye Movement (NREM).
2. Rapid Eye Movement (REM).

1.3.1 NREM Sleep

This represents the first part of the sleep cycle, and it is subdivided into three distinct stages: stage I, stage II, and stage III (also known as N1, N2, N3). Muscle paralysis does not occur appropriately during NREM sleep; that's why some people do not progress to REM sleepwalk during their sleeping because they do not lose their motor function; also, dreaming rarely occurs in this type (Tubbs et al. 2019).

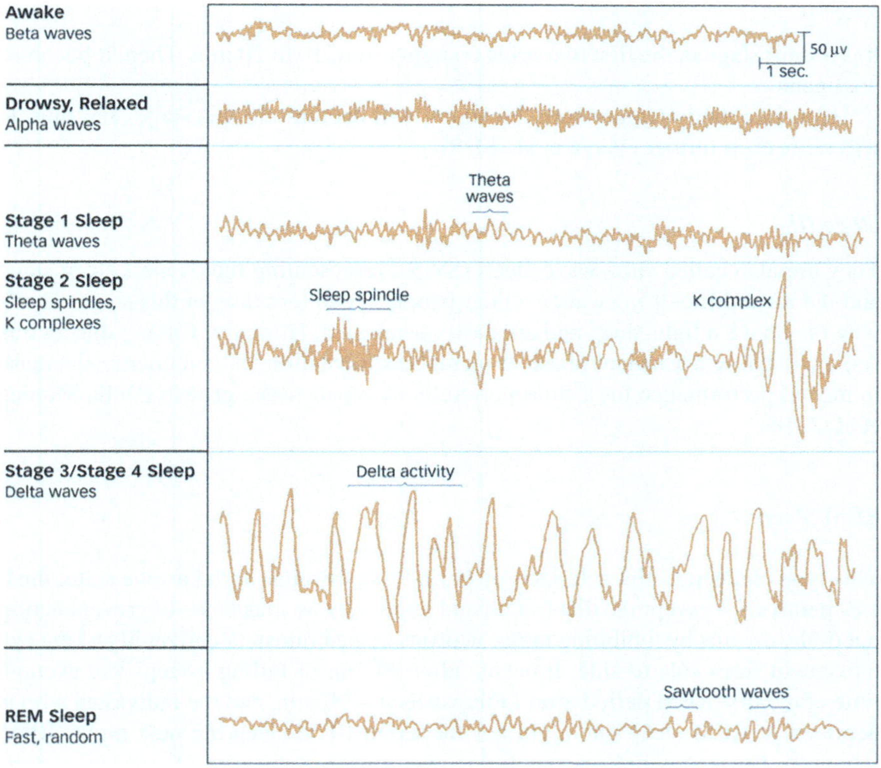

Fig. 2 NREM and REM sleep

Stages of NREM Sleep

Stage I

Sleep starts by a transition from wakefulness to drowsiness state, during wakefulness—when eyes are open—alpha and beta waves are present, and beta waves are the predominant ones, while during the drowsiness state—when eyes closed—alpha waves (with frequency 8–12 Hz) become predominant. Stage I starts when these alpha waves are replaced by theta waves (with frequency 4–7 Hz). During this stage, a selective arousal threshold will determine if a specific minor stimulus is worth responding to and trigger the wakefulness state or should ignore it and proceed to the next stage of sleep. Typically, this stage lasts for 5–10 min only, and it's the entry point for stage II.

Stage II

It is a short stage in the first two cycles ranging from 10 to 20 min. Then, it becomes dominant.

Heart rate, body temperature, breathing will decrease in this stage, and muscle will relax even further (Varga et al. 2018).

Stage III

They are also called slow-wave sleep (SWS), representing the deepest sleep stage and the most difficult to awaken. Older people spend less time in this stage. That's why they have a light sleep and are easily awakened. However, waking during this stage will cause a transient phase of mental disorientation and a moderate decrease in mental performance for a short period: body repair tissue growth (Della Monica et al. 2018).

REM Sleep

This type of sleep is characterized by brain waves similar to the awake state, muscles paralysis—except the diaphragm and upper airway muscles—to prevent acting out of the dreams by inhibiting motor neurons in the brainstem and rapid jerking eye movement from side to side. It occurs after 90 min of falling asleep. The average time of the first REM period after falling asleep is 90 min, and the individual who is deprived of REM sleep one night has increased REM sleep the next night (REM rebound). Dreams mainly occur during this type and are free of sensory experiences, visual content, and emotional reasoning. So, REM dreams play a role in memory consolidation and emotional processing of complex events. Time spent in REM sleep decreases with typical aging.

1.4 Sleep Disorders

Poor or insufficient sleep is associated with a wide variety of disturbances in most body systems, including endocrine, metabolic, and nervous system disorders. Now we will go through most of them and explain the main characteristics (Léger et al. 2008).

1.4.1 Insomnia

Approximately 30–35% of adults experience transient insomnia at some point in their lives. The diagnosis is made when these difficulties are reported for at least three nights per week and more than 3 months.

Insomnia is further subdivided into primary insomnia, which is not attributed to any other medical or psychological conditions, and secondary insomnia, which may be associated with other conditions like psychological stress, chronic pain, restless leg syndrome, drugs, and medications (Morin et al. 2006; Harvey et al. 2014).

1.4.2 Circadian Rhythm Sleep-Wake Disorders

Many disorders emerge under this title. The ICSD-3 classifies the circadian rhythm sleep-wake disorders into seven types which include:

1. **Delayed Sleep-Wake Phase Disorder:** There is difficulty sleeping and waking up (more than a 2-h delay in sleep period) and later than normal individuals. This leads the patient to delay and poor performance at work or school due to daytime sleepiness.
2. **Irregular Sleep-Wake Rhythm Disorder:** Characterized by disorganized sleep and wake patterns. It's more observed in older adults and patients with neurodegenerative disease (Spicuzza et al. 2015).
3. **Non-24-h Sleep-Wake Rhythm Disorder:** Also called free-running disorder and characterized by a gradual delay of sleep-onset time from 1 day to the next, so the individual begins to sleep during the daytime hours and then drift back into the night due to failure of the circadian system to entrain to the 24-h day. Mainly occur in totally blind people (Baumann et al. 2014).
4. **Shift Work Disorder:** Occurs to individuals who work on night shifts due to misadjustment of body circadian rhythm to work schedules lead to drowsiness during shift work difficulty falling asleep during the day.
5. **Jet Lag Disorder:** Occurs in individuals who travel to regions with different time zones because the body can't reset its circadian time to the new time zone upon arrival, so it takes some time to correct this (Rosenberg and Van Hout 2014).
6. **Circadian Sleep-Wake Disorder not Otherwise Specified:** Occurs secondary to medical or neurological disorders, for example, dementia, movement disorders, and blindness (Molaie and Deutsch 1997; Salminen and Winkelmann 2018).

1.4.3 Sleep-Related Breathing Disorders

They are abnormalities of respiration that occur during sleep often associated with a wide variety of comorbidities. The degree of airway narrowing can range from snoring to complete collapse of the airway and cessation of airflow. According to ICSD-3, they are classified into four types: obstructive sleep apnea (OSA), central sleep apnea (CSA), sleep-related hypoventilation, and sleep-related hypoxemia disorder (Maurer et al. 2010).

1. **Obstructive Sleep Apnea (OSA):** One of the diseases commonly found among patients relates to the respiratory system (Fig. 3). Usually, OSA is accompanied by snoring and individuals unaware of their breathing difficulty even when

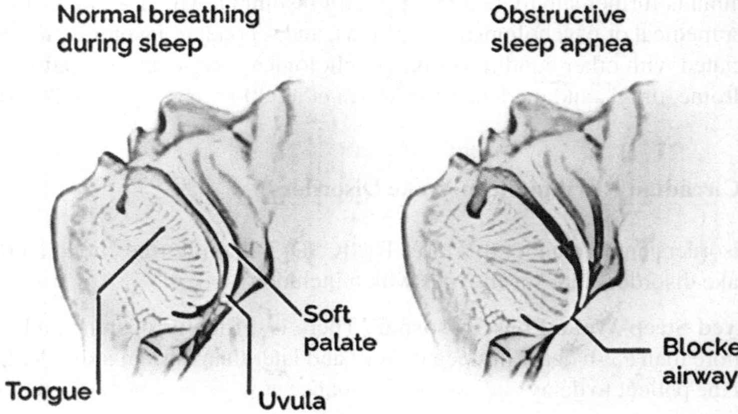

Fig. 3 Mechanism of obstructive sleep apnea

walking at night until their sleep partner or others recognize it. These events occur primarily during sleep stages I, II, and REM and are associated with severe desaturation, while sleep stage III is protective against OSA with less severe desaturation. In addition, individuals may have symptoms of unexplained daytime sleepiness, restless sleep, morning headache, and mood changes. Causes of OSA may be due to old age, traumatic brain injury, decreased muscle tone, and obesity. The gold standard for OSA treatment is continuous positive airway pressure (CPAP).

2. **Central Sleep Apnea (CSA):** It means that the origin of the pathology comes from the brain that stops ventilation repetitively due to lack of brain signals that drive respiratory muscles to control breathing during sleep, so there is no respiratory effort, and this is in contrast to OSA where respiratory signals from the brain are normal and the problem in the upper airway which obstructs and not open properly. CSA is divided into two categories:
   (a) ***Hypercapnic type:*** In this type, the brain fails to send signals to stimulate respiratory muscles for breathing due to narcotic drugs (e.g., opioids), stroke, or trauma that affects the brainstem and inhibits respiratory signals, or due to neuromuscular disease (amyotrophic lateral sclerosis, multiple sclerosis) which leads to weakness in respiratory muscles then leads to a buildup of carbon dioxide.
   (b) ***Hypocapnic type:*** Occurs because of aberrant pacing and control of respiration that leads to quick deep breath. Treatment of CSA depends on the cause and the patient's comorbidities; however, we can use CPAP and oxygen supplementation. Also there is an implantable device that stimulates breathing muscles.

3. **Sleep-Related Hypoventilation:** Characterized by insufficient ventilation results in accumulation of carbon dioxide and elevation of $PaCO_2$ during sleep.

It may be accompanied by daytime hypoventilation which further worsens the symptoms.

4. **Sleep-Related Hypoxemia:** Low O_2 saturation to <88% during sleep. As a result, individuals may have complications of polycythemia, heart failure, pulmonary hypertension, and cognitive dysfunction (Van der Salm et al. 2014).

1.4.4 Central Disorders of Hypersomnolence

Cataplexy is a sudden loss of muscle tone with full consciousness when exposed to emotional triggers, and it's the main feature that differentiates between both types of narcolepsy sleep fragmentation and can also be seen in narcolepsy type 2; it is atypical for IH.

The loss of orexin neurons occurs due to genetic and environmental factors like autoimmune attack and destruction in susceptible individuals. Individuals with narcolepsy pass abruptly from waking to REM sleep with a bit of or absent non-REM period.

Idiopathic hypersomnia (IH) is characterized by long non-refreshing naps with/without long sleep time. The main mechanism is unknown and usually doesn't respond to normal treatment used in narcolepsy.

Treatment includes a combination of pharmacological and behavioral therapy. Modafinil is the drug of choice for narcolepsy; it improves wakefulness by reducing the reuptake of dopamine to decrease excessive daytime sleepiness. Venlafaxine is a serotonin-norepinephrine reuptake inhibitor and antidepressant agent used to reduce cataplexy. Usually, to get a maximum effect of these medications, we use them with lifestyle modifications like advising the patient to take scheduled short naps during the day and maintain a regular bedtime. This will reduce daytime sleepiness. However, these naps are ineffective in patients with IH. We also need to treat other comorbidities that accompany narcolepsy because people with narcolepsy tend to be obese, which may affect their lives. Safety precautions should be taken before doing some activities like driving cars and working in jobs that require full alertness during the day because patients with narcolepsy can have serious injuries or death if they fall asleep.

1.4.5 Parasomnia

It means abnormal movements and actions occur during sleep (before sleep, during sleep, or at arousal period after sleep) (Stefani and Högl 2019).

Parasomnias Associated with NREM Sleep

This parasomnia group is associated with arousal and occurs during stage III of NREM sleep. In addition, certain triggers can induce this type of parasomnias like alcohol, sleep deprivation, physical activity, emotional stress, depression, and certain medications; this type includes:

1. **Confusional Arousal:** It's a brief episode of arousal occurring when sleep is interrupted, and the individual is awakened during the first half of the night (during stage III) and characterized by mental confusion, disorientation, amnesia of the event; that's why their sleep bed partner may only note the episode, and the individual can't understand what's going on. Most commonly occur in children and resolve by age 5, in rare cases may continue to adulthood.
2. **Sleepwalking (Somnambulism):** A series of complex behaviors that occur during sleep and ambulation. The individual looks awake with open eyes and performs inappropriate actions like moving around, but actually, they are asleep. This may lead to serious injuries because individuals are unaware of what they are doing due to altered consciousness and impaired judgment. This episode may terminate when the individual returns to bed or lies down outside and continues their sleeping. The exact cause of sleepwalking is unknown, but some studies show it runs in families.
3. **Sleep Terrors:** Patients wake up suddenly and start screaming and maybe crying as well, typically lasting between 30 s and 5 min. The affected individual appears confused, diaphoretic, tachypneic, tachycardiac, and may sleepwalk during attacks, and usually, they don't remember the event the following day. It is challenging to communicate and console the individual during the attack. Individuals with sleep terrors run or jump around in an attempt to avoid harm, thereby injuring themselves or others.
4. **Sleep-Related Eating Disorder:** The exact underlying mechanism of sleep-related eating disorder is unclear, but it is associated with other sleep disorders like sleepwalking, obstructive sleep apnea, narcolepsy, and periodic limb movement disorder; also some medications (e.g., zolpidem) may trigger it.

Parasomnia Associated with REM Sleep

It includes REM sleep behavior disorder (RBD), recurrent isolated sleep paralysis, and nightmare disorder.

1. **REM Sleep Behavior Disorder (RBD):** Recurrent episodes of verbal or complex motor behavior during REM sleep. Because the dream content is violent, the patient may exhibit jerky limb movement, punching, shouting, screaming, hitting, and running, resulting in serious injury to the patient and their bed partner. In addition, the patient may recall the events and dream content after waking up.

2. **Recurrent Isolated Sleep Paralysis:** Loss of voluntary muscle movement at sleep onset or upon awakening. The patient is conscious and may feel unable to breathe; although the diaphragmatic muscle function is unaffected, this may be due to atonia of accessory respiratory muscles. During the episodes, the patient may experience visual, auditory, or tactile hallucinations and usually resolve after seconds or minutes and can be aborted spontaneously by touching or speaking to the affected patient. Sleep deprivation, irregular sleep-wake schedule, stress, and sleeping in a supine position may trigger the episodes.
3. **Nightmare Disorder:** In contrast to other sleep parasomnia, patients with nightmares recall the details of the dream content. Nightmares commonly occur following acute stress disorder and post-traumatic stress disorder (PTSD) and result in episodes of fear, terror, and anxiety accompanied by increased heart rate and respiratory rate. Certain medications like antihypertensive can induce nightmares (e.g., beta-blockers), antidepressants (e.g., SSRI), dopamine agonists, acetylcholinesterase inhibitors, and alcohol.

Other Parasomnias

This category of parasomnia disorders can occur in both NREM and REM sleep as well as during the transition between sleep and wakefulness. They include:

1. **Exploding Head Syndrome:** A loud sound heard in the head upon awakening.
2. **Sleep-Related Dissociative Disorder:** Characterized by dissociative episodes that occur just before the sleep that last minutes to hours and mimic other parasomnias; however, it always occurs during wakefulness state. It may be accompanied with daytime dissociative symptoms and go away with treatment of a dissociative disorder. Patients may be agitated and violent during the episode and result in self-injury.
3. **Sleep-Related Hallucinations:** In this condition, the patient experiences auditory, visual, tactile, and kinetic hallucinations at sleep onset or upon waking. The patient wakes up terrified and may jump out of bed and get injured. This hallucination episode may last for a few minutes and go away when the light is turned on.

1.4.6 Sleep-Related Movement Disorders

They are a wide range of disorders explained one by one as in the following:

1. **Restless Legs Syndrome (RLS):** Also known as Willis-Ekbom disease, which is a chronic sensorimotor disorder characterized by an irresistible urge to move the legs accompanied by an unpleasant sensation in the legs. To ensure that the case is RLS, we must exclude other conditions that explain the symptoms like sleep-related leg cramps, positions, discomfort, habitual foot tapping, akathisia, and arthralgia. RLS can be idiopathic or secondary. Idiopathic form usually begins slowly before 40 years of age, and the patient may have a family history

of RLS, while in secondary form, it has a later onset in life and is associated with other conditions like neurological disorders (e.g., multiple sclerosis), end-stage kidney disease, iron deficiency, or pregnancy. Dopamine dysfunction may have an essential role in the pathophysiology of RLS. Although many patients show significant improvement after administration of dopaminergic agents, recent studies show hyperdopaminergic states in RLS patients, in contrast to the hypodopaminergic state, which is thought to be the primary mechanism for this disorder. Other non-pharmacological treatments like sleep hygiene, pneumatic pressure therapy, exercise, massage, and hot baths may be helpful to reduce the symptoms (Stumbrys et al. 2012).

2. **Periodic Limb Movement Disorder (PLMD):** The contractions last 0.5–10 s, and each episode consists of at least four consecutive movements, with 5–90 s intervals in between. PLMD occurs in the first half of the night during NREM sleep. PLMD is present in about 80–90% of RLS patients. Dopaminergic region dysfunction within the hypothalamus and impaired iron availability in the brain may be involved in PLMD.
3. **Sleep-Related Leg Cramps:** Intense, short-lived painful contractions of calf or foot muscles relieved by stretching, messages, and heat application on the affected muscles. Most commonly occur at night or during the sleep-waking of the patient. Older adults are most commonly affected. Although sleep-related leg cramps are more prevalent among elderly people, they may result from neuromuscular disorders (e.g., radiculopathies, myopathies, and Parkinson disease), electrolyte disturbances, and medications (e.g., long-acting beta) agonists and thiazide diuretics) and could be idiopathic with no relation to other diseases. One-half of patients with sleep-related leg cramps have several episodes per week or day. This may lead to sleep disturbances, difficulty falling asleep, awakening at night and excess daytime sleeping.
4. **Sleep-Related Bruxism:** Polysomnography and masseter electromyography are used to confirm the diagnosis. Sleep bruxism can result in morning jaw muscle pain, temporal headache, tooth destruction, masticatory muscles hypertrophy, and temporomandibular joint discomfort. Unfortunately, there is no effective treatment that cures or stops sleep bruxism. However, oral appliances can be used to protect the tooth from damage.
5. **Sleep-Related Rhythmic Movement Disorder (SRRMD):** Rhythmic, stereotyped with large-amplitude and low-frequency (0.5–2 Hz) body movements that involve large muscle groups and occur before falling asleep or during sleep (particularly stage II NREM sleep). Movement episodes last for a few minutes and manifest as body rocking, headbanging, and head rolling. Severe cases of SRRMD can be treated with benzodiazepines such as clonazepam.
6. **Benign Sleep Myoclonus of Infancy (BSMI):** It is a benign self-limiting disorder characterized by repetitive flexion, extension, abduction, and adduction myoclonic jerks during the NREM sleep period that last for a few seconds and disappear with arousal or movement. Myoclonus jerks are more prominent in upper than lower extremities and begin during the first month of life (first

observed at a median age of 3 days) and go before 6 months. Neurological assessments of the infant are typically normal (Stumbrys et al. 2012).

7. **Propriospinal Myoclonus at Sleep Onset (PSM)**: Repetitive axial jerky movements during the sleep-wake transition period result in sleep-onset insomnia. These movements generate in the spinal cord and spread rostrally and caudally via propriospinal pathways. It is important to do polysomnography with multichannel surface electromyography (EMG) to diagnose PSM. Few cases of PSM occur due to structural lesions in myelin, but the majority are due to functional movement disorders.

2 Dreams

Many cultures worldwide suggest that dreams are a gateway to communicate with other worlds and supernatural entities (Baird et al. 2019).

Throughout their theories, Sigmund Freud and Carl Jung put the first step about dreams interpretations. Sigmund Freud hypothesizes that our subconscious can reveal the wishes that used to be repressed by our conscious mind, so, during dreams, we live out our deepest wishes and desires as he stated that dreams are the "royal road to the unconscious." Carl Jung had another view on dreams that, in contrast to Freud view he believes dreams are attempts to lead the individual toward wholeness through dialog between the ego and the self (ego represents our conscious mind while self represents the totality of our physical, biological, psychological, social, and cultural being that involve the conscious as well as the unconscious).

Dreams mainly occur during REM sleep, so at this stage, circuits in the brainstem are activated then trigger areas of the limbic system responsible for emotions, sensations, and memories, including the amygdala and hippocampus. Hence, the brain tries to interpret this electrical activity to create meaning from these signals.

In cognitive neuroscience, the present theory assumes that dreams have a role in memory consolidation and long-term memory enhancement. Other theories suggest that dreams have a role in controlling emotions and resolving the problems that occur in our daily lives (Baird et al. 2019).

2.1 Dream Properties and Content

Dreams are characterized by multimodal perceptual content, dominated mainly by visual and auditory elements, while smell, taste, movement, and tactile sensations are less frequently experienced. Emotional content is an important feature of a dream; the dreamer mostly experiences negative emotions like fear or anxiety. Sometimes dreams appear strange, unrealistic, or fantasy in content compared to waking state. This "bizarreness" of dreams could be classified into two types:

temporal discontinuity when the scene changes suddenly with little or no transition in between, for example, when the dream starts at a specific place (e.g., at home) and suddenly change to an unrelated location (e.g., in a ship). The other is the unlikely combination where two or more dream elements are unlikely to be combined at the same time according to waking experience (e.g., playing golf on the plane). These bizarre events are taken seriously as real-life experiences by the sleeper as they are unaware this is not real and just a dream until waking up (except in lucid dreams).

Memory recall of dreams at awakening is poor; that's why we forget the exact elements of the dream. Some individuals may have difficulty in a verbal description of their dreams due to inherent bizarreness. Assessment of dreams scientifically is not easy because the dreamer is the only observer of his dream, so a third observer can't access any subjective experience they pass through. The only way to obtain the dream information depends on personal memory recall (Baird et al. 2019).

2.2 Dreams from Neurobiological Perspective

During REM sleep, the cholinergic neurons in pedunculopontine tegmentum (PPT) and laterodorsal tegmentum (LDT) are activated while serotonergic neurons in the raphe nucleus and noradrenergic neurons in locus coeruleus are repressed; this leads to an imbalance between these opposed neurons then leads to activation of the brain during this stage. At the same time, both sensory inputs and motor outputs are inhibited. Sensory suppression is caused by inhibition of presynaptic afferent fibers from sensory organs and muscles to the spinal cord, while motor suppression caused by inhibition of postsynaptic neurons of the anterior horn cells. When an individual has visual and auditory perceptions during REM sleep, this is due to endogenous activation of occipitotemporal cortical areas (inferior temporal cortex and fusiform gyrus both belong to visual association area), so lesions in this area lead to loss of these types of perceptions.

The primary visual cortex functionally separated from the visual association area during REM sleep leads to endogenous activation of the visual system away from the external environment.

Amygdala is one of two almond-shaped structures that accounts as a part of the limbic system located within the temporal lobes, medial to the hypothalamus and adjacent to the hippocampus and inferior horn of the lateral ventricle, typically amygdala responds to threatening stimuli and modulates the function of the hypothalamus.

Decreasing brain activity in the lateral prefrontal cortex and some areas of the parietal cortex may explain why the dreamer is unable to build logical, realistic, and meaningful dreams because these areas are working in combination and are responsible for high cognitive functions and episodic memory recall during waking state, so suppression of these cortical areas make the dreamer have bizarre events.

2.3 Lucid Dreaming

This type of dreaming is unique because it belongs to the area between dreaming and consciousness, defined as awareness of the dream state and environment with the ability to concentrate and make decisions. The individual can interpret the dream while dreaming and remember it after awakening and control the sequence of the events. The cardinal feature of transparency is the cognitive realization of "This is a dream!" or awakening within the dream. When this realization or awakening occurs, it is called a lucid dream. Although most lucid dreams occur during REM sleep, some studies show they can occur during NREM sleep. The same neurobiological basis of lucid dreams has not been identified yet. However, a hypothesis suggests that lateral prefrontal cortices (normally inactive during REM sleep) will remain active during lucid dreaming (Pigeon and Mellman 2017).

For most people, natural lucid dreams occur infrequently and may never occur in their lifetime. There are two types of lucid dreams; the first, when the individual becomes conscious and aware of their dream during the dream; this type is called "Dream-initiated lucid dreams." The second, when the individual is conscious and awake, then enter lucid dreams, and this type is called "Wake-initiated lucid dreams" (Pigeon and Mellman 2017).

2.4 Dreams, Sleep, and Consciousness Relationship

Consciousness is now considered to have two main aspects. First, consciousness as *awareness* (phenomenological meaning); second, consciousness as *strategic control* (functional meaning). As the phenomenological aspect is the primary concern, it is classified into three main types of awareness.

(a) **Awareness as a phenomenal experience of events:** It means the awareness of recognizing things and events and is sometimes referred to as primary consciousness, which is defined as the experience that results from a response to perception, feelings, thoughts, and memories. In primary consciousness, we are the subjects who do the thinking, feeling, and actions in response to external environments. The target of concern in phenomenal experience may be considered external or from the outside environment. In this condition, the awareness represents an indirect report of the external world (e.g., I can see a car or hear my friend's voice). Or the target of concern may be considered internal or from inside our body, and in this condition, the awareness represents direct reports of the inner world (e.g., I am hungry or I have pain in my arm).

(b) **Awareness as meta-awareness (also called meta-consciousness):** Means awareness of mental life itself or when our consciousness re-represents the contents of everyday experience. During a dream, meta-awareness is expressed in various ways and interpreted as reality testing, which is the mental action when the individual can decide whether an experience is a result of information out of the brain (external world) or whether it is a product of the mind, as in the dream—also referred to as self-reflectiveness, which suggests that the dreamer consciously reflects the truth that they are dreaming. However, reality testing is vague in some situations, so the dreamer can't decide whether this is the result of mental activity or the external world. The dreamer may take part in the event or watch it as an external observer.

(c) **Awareness as self-awareness:** To activate the meta-awareness about the objects, events, feelings, and thinking, we must first be aware of ourselves or our entity. However, self-awareness can be present when meta-awareness is absent.

During the dream, phenomenal awareness is present. In contrast, meta-awareness is supposed to be absent so the individual can neither determine their thoughts that produce the dream nor experience the reality testing, so the dreamer is unaware that they are dreaming. This event is not actual, so this may explain the bizarreness of dreams. When meta-awareness is present and does not fade during a dream, dreamers can recognize that they are dreaming. That occurs in lucid dreams where the dreamer is aware of dreaming and the self as a dreaming subject and has cognitive insight and access to memory and the capability of control on their dream.

Self-awareness represents the conscious state where the individual's attention is oriented toward external events to their reflective consciousness and personal history. This type of awareness is implicated only in the direct experience of perception, sensation, and thoughts. All the operations involved in dream production are unconscious. From the functional point of view, the role of the conscious process in the dream is only the mental activities that monitor the output of the unconscious elaboration, interpretation of what is happening in the dream scene, self-awareness, and emotional experiences. In general, self-regulation and reflective consciousness are inefficient during dreaming. While the primary consciousness and self-awareness remain intact and efficient in the oneiric world, meta-awareness is not always preserved (Pigeon and Mellman 2017).

2.5 *Dreaming About Future and Postconscious*

Dreaming about one's future events and thoughts does not take enough time to be saved in our consciousness. That's why our postconsciousness releases them to make them satisfied and comfortable about the frightens. The dreams that cause you to be comfortable, which I will call good dreams, while others make you uncomfortable, and I will call them bad dreams. Postconsciousness creates these dreams to release the thoughts from our mind and to make us more tolerated and realistic with our reality, not with our imaginations, so it creates some kind of imaginations (could be real or not and could be good or bad), to fight our worries and stressing thoughts.

Multiple Choice Questions

1. Which stage of sleep is characterized by the (WHS Nervous System Flashcards. https://www.flashcardmachine.com/whs-nervous-system.html) presence of sleep spindles and K-complex with the decrease of heart rate, respiratory rate, and body temperature:
   (a) Stage I of NREM
   (b) Stage II of NREM
   (c) Stage III of NREM
   (d) REM sleep
   (e) None of the above

2. The minimal sleep requirement for a 4-year-old child is:
   (a) 9
   (b) 7
   (c) 10
   (d) 14
   (e) 11

3. The primary structure that is responsible for controlling our sleeping and waking cycle is
   (a) Suprachiasmatic nucleus
   (b) Medulla
   (c) Thalamus
   (d) Pineal gland
   (e) Amygdala

4. Insomnia characterized by all of the following features, except:
   (a) Dissatisfaction with sleep quality and quantity
   (b) Disturbance of sleep can be at sleep onset or/and sleep maintenance
   (c) May associated with psychological stress, chronic pain, restless leg syndrome, and drugs
   (d) Decrease in GABA neurotransmitters in the brain
   (e) It occurs *3 times:1 week* during at least 1 month

5. One of the following is NOT a feature of circadian rhythm disorder:
   (a) Difficulty in falling asleep and difficulty in waking up at the desired time mostly occurs in people with neurodegenerative disorders
   (b) Multiple short periods of sleep and wakes through the day with normal sleep duration
   (c) Jet lag disorder occurs in people who travel to areas with different time zones
   (d) May occur due to dementia and blindness
   (e) Usually, they are treated by timed blue light and melatonin

6. In obstructive sleep apnea all the following are characteristics, except:
   (a) Repetitive episode of upper airway obstruction due to relaxation of throat muscles
   (b) Occur during sleep stage III and REM and associated with severe desaturation
   (c) Associated with daytime sleepiness, mood changes, and morning headache
   (d) May occur due to old age, traumatic brain injury, and obesity
   (e) Best treatment option is continuous positive airway pressure
7. All of the following are parasomnia disorders associated with NREM sleep, except:
   (a) Sleep walking
   (b) Sleep-related eating disorder
   (c) Nightmare disorder
   (d) Sleep terrors
   (e) Confusional arousal
8. A 33-year-old female complains of an irresistible urge to move her legs that start after she finishes her work that increases at night and relieved by moving. She denied any short time painful leg contraction, other neuromuscular disorder, or joint pain. These features mostly indicate that she has:
   (a) Sleep-related leg cramps
   (b) Sleep-related rhythmic movement disorder
   (c) Restless leg syndrome
   (d) Periodic limb movement disorder
   (e) Propriospinal myoclonus of sleep onset
9. Activation-synthesis hypothesis of the neurobiological theory proposed that dreams are:
   (a) Imaginary visions and hallucinations
   (b) A way of connection between our world and God's world
   (c) Electrical brain signals that activate certain areas of the brain and pick random thoughts, images, and feelings of our memory
   (d) Training places to improve our skills to avoid real life-threatening situations
   (e) Have an important role in long-term memory consolidation and enhancement
10. One of the following is NOT a feature of lucid dreaming:
   (a) Belong to the area between dreaming and consciousness where the individual is able to make decisions and changes in his or her dream
   (b) May occur due to persistence activation of lateral prefrontal cortex
   (c) Mostly it occurs during REM sleep period
   (d) It can be induced using different strategies such as cognitive technique and external stimulation
   (e) Absent of meta-awareness and self-awareness

Answers

1. (b)
2. (e)
3. (a)
4. (e)
5. (a)
6. (b)
7. (c)
8. (c)
9. (c)
10. (e)

References

Baird B, Erlacher D, Czisch M, Spoormaker VI, Dresler M. Consciousness and meta-consciousness during sleep. In: Handbook of behavioral neuroscience, vol. 30. Amsterdam: Elsevier; 2019. p. 283–95.

Baumann CR, Mignot E, Lammers GJ, Overeem S, Arnulf I, Rye D, Dauvilliers Y, Honda M, Owens JA, Plazzi G, Scammell TE. Challenges in diagnosing narcolepsy without cataplexy: a consensus statement. Sleep. 2014;37(6):1035–42.

Della Monica C, Johnsen S, Atzori G, Groeger JA, Dijk DJ. Rapid eye movement sleep, sleep continuity and slow wave sleep as predictors of cognition, mood, and subjective sleep quality in healthy men and women, aged 20–84 years. Front Psych. 2018;9:255.

Harvey AG, Bélanger L, Talbot L, Eidelman P, Beaulieu-Bonneau S, Fortier-Brochu É, Ivers H, Lamy M, Hein K, Soehner AM, Mérette C. Comparative efficacy of behavior therapy, cognitive therapy, and cognitive behavior therapy for chronic insomnia: a randomized controlled trial. J Consult Clin Psychol. 2014;82(4):670.

Léger D, Poursain B, Neubauer D, Uchiyama M. An international survey of sleeping problems in the general population. Curr Med Res Opin. 2008;24(1):307–17.

Maurer VO, Rizzi M, Bianchetti MG, Ramelli GP. Benign neonatal sleep myoclonus: a review of the literature. Pediatrics. 2010;125(4):e919–24.

Molaie M, Deutsch GK. Psychogenic events presenting as parasomnia. Sleep. 1997;20(6):402–5.

Morin CM, Bootzin RR, Buysse DJ, Edinger JD, Espie CA, Lichstein KL. Psychological and behavioral treatment of insomnia: update of the recent evidence (1998–2004). Sleep. 2006;29(11):1398–414.

Pigeon WR, Mellman TA. Dreams and nightmares in posttraumatic stress disorder. In: Principles and practice of sleep medicine. Amsterdam: Elsevier; 2017. p. 561–6.

Roenneberg T, Kantermann T, Juda M, Vetter C, Allebrandt KV. Light and the human circadian clock. Handb Exp Pharmacol. 2013;(217):311–31.

Rosenberg RS, Van Hout S. The American Academy of sleep medicine inter-scorer reliability program: respiratory events. J Clin Sleep Med. 2014;10(4):447–54.

Salminen AV, Winkelmann J. Restless legs syndrome and other movement disorders of sleep—treatment update. Curr Treat Options Neurol. 2018;20(12):1–2.

Schupp M, Hanning CD. Physiology of sleep. BJA CEPD Rev. 2003;3(3):69–74.

Sollars PJ, Pickard GE. The neurobiology of circadian rhythms. Psychiatr Clin. 2015;38(4):645–65.

Spicuzza L, Caruso D, Di Maria G. Obstructive sleep apnoea syndrome and its management. Ther Adv Chronic Dis. 2015;6(5):273–85.

Stefani A, Högl B. Diagnostic criteria, differential diagnosis, and treatment of minor motor activity and less well-known movement disorders of sleep. Curr Treat Options Neurol. 2019;21(1):1.
Stumbrys T, Erlacher D, Schädlich M, Schredl M. Induction of lucid dreams: a systematic review of evidence. Conscious Cogn. 2012;21(3):1456–75.
Tubbs AS, Dollish HK, Fernandez F, Grandner MA. The basics of sleep physiology and behavior. In: Sleep and health. Cambridge: Academic Press; 2019. p. 3–10.
Van der Salm SM, Erro R, Cordivari C, Edwards MJ, Koelman JH, van den Ende T, Bhatia KP, van Rootselaar AF, Brown P, Tijssen MA. Propriospinal myoclonus: clinical reappraisal and review of literature. Neurology. 2014;83(20):1862–70.
Varga B, Gergely A, Galambos Á, Kis A. Heart rate and heart rate variability during sleep in family dogs (Canis familiaris). Moderate effect of pre-sleep emotions. Animals. 2018;8(7):107.

Defense Mechanisms and Personality Disorders

Saleh Abdulkareem Saleh

1 Defense Mechanism

In the nineteenth century, Sigmund Freud (1856–1939) exhibited his ideas about the defense mechanisms and their relation to the human mind, which is represented by the Id (unconscious fundamental and unorganized part of our mind that represent the main drive of our needs, desires, and impulses that seek for immediate pleasures), ego (present in the conscious and unconscious part of our mind responsible for organization and planning, and manifest by the defensive, reasoning, reality testing, intellectual and executive functions that regulate between our primitive demand of Id and the higher moral values of the superego), and superego (present in the conscious, preconscious, and unconscious part of our mind that symbolize the higher moral values and principles that associate with the society rules, so it guides our behavior to what is right and wrong and function by suppressing the unacceptable Id impulses). After that, in the twentieth century, his daughter Anna Freud tried to complete his work by defining, analyzing, and adding more of these defense mechanisms. Also, she proposed that defense mechanisms depend on the maturity of the ego. Defense mechanisms are mental processes that act unconsciously to resolve the conflicts that result from rapid change in internal and external reality through distortion or manipulation to avoid suffering from anxiety and depression. These conflicts may interfere with one's family, culture, or identity values. Which defense mechanism takes the act is something that we cannot control, but it has many consequences on our mental health. When the individual uses the defense mechanisms as a main coping style with life stressors, this may damage the self and result in irrational, harmful behaviors. So defense mechanisms act as a way of masking reality and preventing the individual from facing the problem and taking responsibility. The target of the defenses is to rebalance the psychological state and

S. A. Saleh (✉)
University of Baghdad, College of Medicine, Baghdad, Iraq

H. T. Hashim, A. Alexiou (eds.), *The Psychology of Consciousness: Theory and Practice*, https://doi.org/10.1007/978-3-030-90692-4_5

provide the mind with time to adjust to the new changes. Also, it leads to a decrease in cognitive, emotional, and physiological stress. The motives of these defenses are to ward off the instinctual anxiety feeling of guilt and protection of our self-esteem. Defense mechanisms work through alteration of perception and relationship between self, others, and the conflicts. Although defense mechanisms lead to psychological adjustment and more adaptation to the internal and external reality, sometimes it may cause psychological distortion and result in pathology. These defenses have an important role in psychodynamic therapy; it helps make the individuals more aware of their behaviors and allow them to understand themselves more deeply. So it could be a valuable tool to manage people with depression, anxiety, and personality disorders.

2 Classification of Defense Mechanisms

There are many theories developed to classify the defense mechanisms of which the borderline personality organization theory that developed by Otta Friedmann Kernberg in 1967 and depend on the object relation theory to explain how the personality develops during childhood using the primitive defense mechanisms (e.g., projection, denial, dissociation, and splitting) as a cornerstone of this theory. Another theory developed by Robert Plutchik (1927–2006) in 1979 depends on the eight opposite primary emotions to explain our defense mechanisms, for example, fear represents repression, anger represents displacement, joy represents reaction formation, sadness represents sublimation, acceptance represents denial, disgust represents projection, expectation represents intellectualization, and lastly surprise represents regression, and all the other emotions and defenses are derived from these. Finally, the last theory developed by the psychiatrist George Eman Vaillant claims that defenses are a continuation to their psychoanalytic developmental level and classify the defense mechanisms into a hierarchical model of four main categories ranging from the pathological level of defenses to the healthiest adaptive level (Fig. 1). This is the same classification used by the American Psychiatric Association (ASA) through their (DSM-4) diagnostic and statistical manual for mental disorders (Fig. 2). We will go through most of the defense mechanisms according to Vaillant's

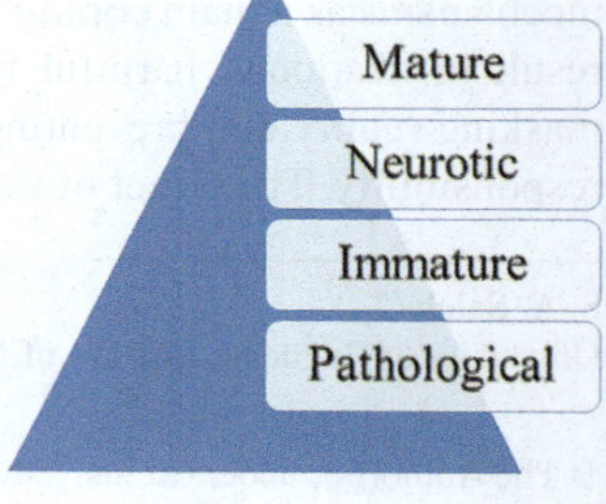

Fig. 1 Vaillant classification of defense mechanisms level from least adaptive (pathological) to highest adaptive (mature)

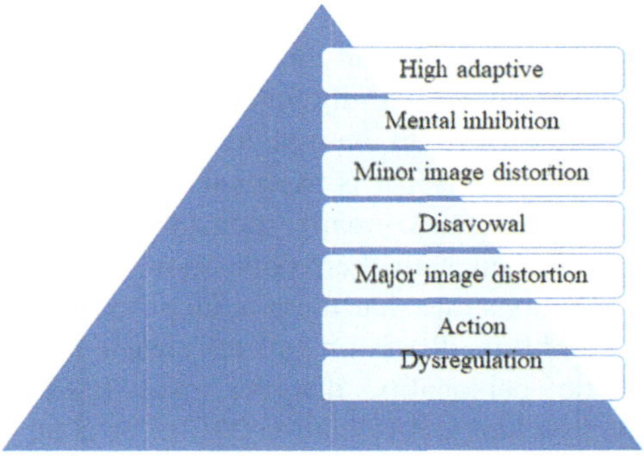

Fig. 2 Diagnostic and statistical manual, fourth edition (DSM-4), classify the defense mechanisms into seven levels from the least adaptive (dysregulation level) to the highest level (high adaptive level)

classification, including more defenses that were discovered after Vaillant's original classification:

2.1 Pathological Defenses

This level represents a failure of the individual defensive system to arrange the response to the stressors, so when these kinds of defenses take the actions, they will result in distortion of external reality to eliminate the individual's need to deal with it and make them think inconsequential and behave in irrational ways. Although they present in some psychiatric disorders (e.g., schizophrenia), they may typically appear in dreams (Vos and De Haes 2007).

1. **Psychotic denial**: Characterized by avoiding the awareness or inability of the individual to accept part of the internal reality (their feeling, thoughts, and experience) and part of the external reality (the outside events and stressors) as it is disturbing then make them to deal and act as if they did not exist despite the overwhelming evidence. For example, an older person who is newly diagnosed with cancer refuses to acknowledge his disease and start treatment despite all the signs and symptoms he suffered from, just because the idea of having cancer is too annoying and very hard to handle. Hence, it is easiest to deny the actual disease.
2. **Delusional projection:** Characterized by fixed beliefs about the external reality or own self that cannot be changed even in the presence of reasonable conflicting evidence regarding its truth. These beliefs may be based on wrong or misleading information. Their content includes many different themes: persecutory (the most common type), referential, somatic, grandiose, and religious. It may occur either when the individual perceives their feeling in another individual and acts on that perception, or inversely when they perceive another individual or their feelings inside themselves. Also, it appears in psychosis and other psychiatric disorders. For example, an older man attacks a security guard because he believes the government agency wants to kill him.

3. **Distortion**: Unrealistic extreme reshaping of the external world to meet inner core thoughts and impulses. The unconscious mind focuses only on things that correspond with the internal wishes and ignores all the evidence that disagrees with them. So the individual has automated thoughts that experience during all events. Also, it is associated with unrealistic beliefs, hallucinations, delusion, and denial to own responsibility to take action. Uncomfortable impulses are replaced with their opposition to relieve the internal conflicts. Some studies demonstrate that there is an association between distortion and depression. In addition, distortion appears clearly in individuals with schizophrenia, narcissistic personality disorder, and borderline personality disorder (American Psychiatric Association 2013; Lee et al. 2020).

2.2 *Immature Defenses*

This defense level is responsible for suppressing emotional awareness to minimize the distress and conflict from the external reality. Usually, it is present in adults; however, it can appear in some personality disorders and major depressive disorder. In addition, overuse of these defenses can result in maladaptive behaviors and impair the individual's ability to cope efficiently, which include:

1. **Acting out** is characterized by the direct expression of unconscious feelings and thoughts through physical actions without awareness of the emotions that drive these behaviors. It occurs due to the inability to deal with these mental contents thus cannot express them verbally. Synonymously, acting out is used to refer to impulsive control disorders. It also shows up in patients with borderline or antisocial personality disorders, chronic drug abusers, failure perversion, or self-inflicted injury to relieve the stress. For example, a child with a sick dad misbehaves in school.
2. **Passive aggression** is characterized by indirect or ineffective expression of hostility to others in a non-confrontational manner. This defense mechanism is very familiar, and we use it in many situations when we are angry or dissatisfied with the results. Hence, we indirectly express our anger to let the person in charge know we disagree. This behavior will lead to failures, procrastination, decreased productivity, and can't get things done, leading to a person we can't depend on. Passive aggression is also associated with psychiatric disorders like acute stress disorder, adjustment disorder, borderline personality disorder, and deliberate self-harm. For example, an employee who gets to work late intentionally due to a conflict with his boss (Santana et al. 2017).
3. **Projection**: Characterized by attributing an individual's own unacceptable internal feelings, thoughts, or impulses to an external environment (other person or object), which occur because the individual thinks their thoughts are unacceptable to reveal them directly, so they try to project them on others. During this process, the individual is unaware that these impulses and feelings originate

internally. However, it may appear in children and be considered normal to help them cope with behavioral problems like fears, sadness, shyness, and oversensitivity. However, it indicates impairment in developmental dealing with the world during adulthood. We could notice projection also in people with schizophrenia, victims of childhood sexual abuse, depression, obsessive-compulsive disorder, pain disorder, and anorexia nervosa. For example, a husband who wants to cheat on his wife blames her for being unfaithful and does not love him (Yazdanshenas Ghazwin et al. 2017).

4. **Schizoid fantasy**: Fantasy represents a wide range of creative mental processes used to express and experience specific wishes and desires that are impossible to have in real life, so it serves as a way to resolve uncomfortable situations. Schizoid fantasy is characterized by a tendency to fall back into one's imagination in order to relieve the present conflicts. For example, it is used to gratify hidden desires for interpersonal relationships and to abort the overexpression of aggressive sexual wishes. It dominates in children and schizoid personality disorder. Or they are experiencing a daydream during a conversation as a way to repel boredom.
5. **Splitting**: Categorizing the self and other individuals on the extremes as either all-good (angels) or all-bad (evils) at different times and according to the situation so the split person does not have in between (black-white thinking pattern) and cannot accept the ambiguity and tolerate that the people could be good and bad simultaneously. The person uses splitting to project and maintain the internal contradictory attitudes and conflicts, but also prevent combination between them. Spitting may also represent different sections within personality and consciousness. Sigmund Freud explains splitting by referring to the trauma when the individual experiences a traumatic event. The mind will repress the uncomfortable feelings. After that, these memories will split off from the consciousness and labeled as bad events and the mind tries to avoid them. So, according to Freud, splitting is a normal part of our personality development during childhood that will incorporate and form the id later on. Splitting appears very clearly in patients with narcissistic and borderline personality disorders. For instance, a patient recognizes the surgeon who does the operation to him as the best surgeon in the world and the kindest one. On the second day, when the surgeon does not come to see him, he redefines him as the worst doctor he has ever had and does not respect his patients (Zanarini et al. 2013).
6. **Idealization**: Exaggerated expressing of perfect thoughts about self, others, and the events while ignoring the unpleasant side of the truth. In idealization, there is a minor image distortion of reality. For example, an adult who recalls childhood memories and admires them and considers them the best moment of his life, on the other hand, ignores the stress and bad events that accompany them.
7. **Devaluation** is the opposite of idealization, where the individual regards the self, others, and situations as totally worthless and flawed. In devolution, the individual focuses on the negative feelings and thoughts and ignores all the other positive sides. When the person experiences idealization and devaluation, they may use a splitting defense mechanism. It typically presents during childhood as

part of the development. When the individual becomes mature and realizes that the people can be both good and bad simultaneously, devaluation will decrease. However, continued devaluation during adulthood may refer to a problem in personality development due to childhood trauma. For instance, a person who thinks he does not deserve a good job and a good life (Thomas et al. 2020; Rnic et al. 2016; Esterberg et al. 2010).

2.3 Neurotic Defenses

This type of defense manifests as severe anxiety and arises when the individual experiences emotional distress, instability, and unconscious struggles to alter the internal feelings, so the individual behaves eccentrically. Due to bad behavior, the person may relieve the internal conflicts, but it will affect their relationships and daily life. It commonly occurs in healthy people and individuals with psychiatric disorders such as schizophrenia, bipolar disorder, borderline personality disorder.

1. **Intellectualization**: By this defense mechanism, the individual uses overthinking to deal with an annoying situation then isolates oneself from the feeling to reduce the harm of uncomfortable emotions. So it looks at the event from a fact, rational and abstract reasoning point of view to avoid provoking anxiety. So, for example, a young boy whose parents died in a car accident tries to investigate and understand how the accident happened instead of showing his sad feelings.
2. **Reaction formation**: This defense mechanism manifests by transforming the uncomfortable wishes, thoughts, and impulses that may be considered dangerous into the opposite behavior unconsciously to relieve anxiety, sexuality, and internal guilt. It keeps the original individual feelings and impulses out of the awareness, so they are only recognized when the person shows a behavior diametrically opposite to what was expected for a particular situation. Also, it is considered transitional defense between the low-level immature (acting out) defenses and high-level mature (altruism) defense mechanisms. Finally, it may count as an adaptive defense because it reduces the destructive behaviors and makes the individual quit the bad habits or do more beneficial behaviors. Some studies show an association between reaction formation and anxiety-related disorders such as panic disorder, anxiety disorder, and agoraphobia. In contrast, others demonstrate that patients with bipolar disorder use reaction formation less frequently. For instance, a well-known Stockholm phenomenon is when the kidnap victim falls in love with the perpetrator to build a bond between them and show affection contrary to what she was feeling at that time.
3. **Dissociation**: Drastic modification of customarily integrated functions of identity, memory, consciousness, and behavior to relieve the emotional distress. It ranges from a minor adjustment in character to complete detachment from reality. Dissociated individuals usually disconnect from time and external environment, and sense of self. So it will lead to amnesia and fragmentation in

personality. It can be noticed in individuals who suffered extreme traumatic events during childhood. In severe cases, the person may display multiple personalities. Although dissociation as defense mechanism is temporary, it can present in more frequent states and appears clearly in individuals with dissociative identity disorder, depersonalization and derealization disorder, dissociative amnesia, and patients with post-traumatic stress conversion disorder, and borderline personality disorder. Some studies suggest that dissociation represents the base to rely upon as a mechanism for other defenses to achieve their goals, such as repression, intellectualization, and splitting. For example, a victim girl exposed to sexual abuse when facing the abuser suddenly becomes numb and disconnected from the environment (Tredget 2001).

4. **Displacement**: Unconscious redirection of the acute feelings and impulses toward another less threatening person or object in an attempt to resolve the internal conflicts. Individuals use displacement because they cannot express their feelings directly due to fearing the consequences. This maladaptive mechanism may cause more harm to the individual due to the destruction of relationships and create more problems. Displacement is seen frequently in patients with panic disorder and individuals with childhood sexual abuse. Evidence demonstrates the association between displacement and emotional eating (eating aggressively to cope with aversive feelings, thoughts, and memories). For instance, an angry and conflicted mother with her husband directs her feelings and emotions toward her children.
5. **Repression**: Unconscious elimination of unacceptable and painful thoughts, memories, and feelings from conscious awareness. The individual does not recall the memories of the traumatic events even if being conscious and aware at that time. It may distort the reality to relieve the anxiety and internal impulses; however, continuous repression of events may provoke anxiety and disturb the feelings. These psychological conflicts may represent a threat and feeling of guilt to self, so the mind represses them unconsciously to protect our self-esteem and other relationships. Although the repressed individual may appear healthy as ordinary people, this is a wrong impression. They may be more distressed and have a higher risk for physical disorders. Also, repression repressed the first cornerstone for Freudian psychoanalytic theory and the development of other defense mechanisms. For example, an individual who suffered childhood sexual abuse does not recall any memories of that event (Iacovino et al. 2014; Salvatore et al. 2012).
6. **Isolation of affect**: Unconsciously separating the internal impulses, thoughts, and feelings from the outside environment. So the individual's mind tries to avoid experiencing the painful feelings associated with bad events. Although it is associated with obsessive-compulsive disorder, it can occur in ordinary people after experiencing traumatic situations. Isolation of affect allows the individual to keep conflicting ideas and behaviors without feelings of guilt, humiliation, and anxiety. For example, a person who describes a murder event that occurred in front of him without showing any emotional responses.

7. **Regression**: Temporary or long term and involuntary turning back in maturational state of ego to an earlier developmental stage in a way of seeking more comfort and safety from a stressful situation. So the individual will behave, act, and feel like a child, immature or in an age-inappropriate manner to face the uncomfortable events. Also, it makes the individual dependent on others to make decisions and be assertive to correlate with a doubtful state. The individual's behavior is widely variable and associated with stage fixation during early development. Such an employee who has stressful work and conflicts with his boss returns to his parents' home and lies in his mother's lap (Clarkin et al. 2021).
8. **Rationalization**: Justification of one's feelings, thoughts, and acting for certain events using plausible explanations that do not represent the events. So the individual tries to replace the threatening situation with a more logical one and make excuses to minimize the anxiety and protect self-esteem and self-concept. This defense mechanism refuses to acknowledge the unpleasant aspect of reality, so it deals with conflicts via lack or minimal level of awareness. Some studies show that rationalization is among the most prevalent defense mechanisms in people with borderline, schizotypal, antisocial, and narcissistic personality disorders (Lee 2017; Frick 2016; Zanarini et al. 2009).
9. **Undoing** is an attempt to give up the conscious awareness of thoughts, emotions, and unacceptable and undesirable impulses and then do the reverse behavior to correct them. So the individual will show the opposite meaning of those impulses to revert the results of his or her act or thoughts and as these events had never occurred. It was suggested that undoing as an intermediate level defense mechanism between mature and immature however it is still a maladaptive process. Undoing is one of the most prominent defense mechanisms in patients with obsessive-compulsive disorder. Also it is used more frequently for panic disorder and social phobia. For instance, after fighting, a husband brings his wife something she has been waiting for a while to apologize indirectly to her since he was aggressive to her.
10. **Somatization**: Redirecting the unpleasant feelings and thoughts the individual has for other people toward the self. Due to the psychological distress, the individual may have psycho-physiological disturbance and show distress as any known disease cannot explain physical symptoms. Usually, the signs and symptoms are vague, do not link to each other, and may involve multiple systems in the body. Some psychiatric disorders represent the main form of somatization, such as body dysmorphic disorder, conversion disorder, hypochondriasis, somatization disorder, pain disorder, undifferentiated somatoform disorder, and somatoform disorder not otherwise specified. Somatization may be mainly symptom-based and represent hypochondriasis or signs-based and the conversion disorder. Hypochondriasis is also sometimes considered a defense mechanism by itself. Otherwise, it is labeled as somatic symptom disorder and illness anxiety disorder. Any slight change of the physical symptom makes the individual believe and consider this as evidence of serious illness. It may result from misperception of body and mind conditions without the actual medical disease. Even after the physician evaluation and reassurance, the hypochondria-

sis may persist, and the individual does not believe there is nothing to be concerned about. Also, it considers a nonverbal, interpersonal connection and help-seeking behavior to cope with stressful life events. Hypochondriasis involves a wide variety of symptoms and psychiatric disorders like pain, body appearance, gastrointestinal, cardiopulmonary, sensory-perceptual issues, personality disorder, neurasthenia, obsessive-compulsive disorder, hysteria, depression, and paranoid psychosis. For example, a young boy with a minor cough may think that he has lung cancer, and he is going to die. While conversion disorder (also named a functional neurological symptom disorder in the DSM-5) is also considered a defense mechanism by itself, it is characterized by converting psychological harm into physical symptoms that lead to impairment of sensory and motor functions. Through this process, the individual unconsciously relieves their anxiety—for example, a woman complained of loss of consciousness after having a fight with her husband.

2.4 *Mature Defenses*

Also, referred to as adaptive defense mechanisms that we can develop overgrowing with life and used by healthy individuals to deal with daily situations to resolve the anxiety and impulses without affecting the individual life. They are associated with emotionally intelligent people, life satisfaction, happiness, psychological maturity, and successful relationships. They are consciously used by a person with the intention to increase feelings of gratification and control over life events. Even if the situation is hard to handle, the reality is kept to itself. This contrasts with immature defenses where the reality is distorted and denied to cope with. These defenses try to value the self, look at the events from a new perspective, and create a new meaning. So this level represents the maximum capability to adapt and handle the stressors.

1. **Humor**: Facing the negative impulses, thoughts, and feelings of events using a sense of humor and comedy. The individual uses humor as a distracting method when exposed to uncomfortable situations or is not ready to deal with them. There are studies that show increasing use of humor enhances psychological well-being, life quality and satisfaction, and interpersonal relationships. For example, a medical student who has a challenging exam tends to make jokes and do funny things to cope with his anxiety (Ronningstam 2011; Caligor et al. 2015; Novais et al. 2015).
2. **Sublimation**: Expression of unacceptable thoughts, impulses, and behavior more constructively and acceptably with changing the aim to suit social norms. So the individual will show their internal conflict through more valuable behavior. It is considered one of the most mature defense mechanisms, and only a small number of people can maintain it regularly. Usually, it expresses sexual and aggressive impulses and transforms this negative energy into more positive

activity. Although it processes subconsciously, we can still intentionally channel our urge into more creative and beneficial things. For instance, a young boy who expresses his aggressiveness throughout practicing boxing or an artist who expresses his desires by his artwork (Nestadt et al. 1990).

3. **Suppression**: Conscious decision to eliminate distressing impulses, thoughts, and desires from conscious awareness in an attempt to remove these conflicts from the present reality and deal with them later on. It lets our attention focus on more important things and defer unnecessary issues to other times. What distinguishes suppression from repression and other defenses is that it operates under our control optionally and consciously. Some studies reveal the correlation between suppression and better psychological adjustment, flexibility, happiness, and adaptation to the environment. While less use of suppression as a defense mechanism is associated with some psychiatric disorders like borderline personality disorder and anxiety disorder, improvement of suppression through behavioral therapy can improve the symptoms of obsessive-compulsive disorder and depression. For example, a student who has an exam tries to suppress all other distractions to focus on his exam (Morrison 1989).
4. **Altruism**: Alleviating negative impulses and emotions by providing services to others to increase self-satisfaction and gratification about oneself. Usually, the help is provided to individuals who have the same feelings and struggles without any obligation or expectation of reciprocal rewards. Some studies show that altruism is associated with better physical, mental well-being, and romantic relationships. For example, a cancer survivor patient makes a large donation to charity for helping cancer patients (Sulz and Hysteria 2010; McGauley et al. 2011).
5. **Anticipation**: The ability of the individual to expect the future events then process them and make realistic planning and preparation for the potential risk or stressful situations that may result from, in an attempt to mitigate the anxiety. It involves either managing the uncomfortable event and putting in the effort to fix it or just looking at the events from another perspective and redefining it as less risky and feeling good about it. In addition, there are certain approaches used to enhance the anticipation for the individuals also, it depends on the interaction between perceptual and cognitive skills. For example, a young man prepares himself for a job interview and plans his answers to any related questions.
6. **Affiliation**: The tendency to belong to other people to resolve and deal with internal and external stressors, share problems without considering other people responsible for them, and make a connection and cooperation. When an individual being with others can provide a chance for distraction and a host of other emotional support. For example, alcoholic individuals participate in group therapy (Shea et al. 2004).

3 Personality Disorders

Personality represents a continuous fixed repetitive pattern of thinking, reacting, feeling, and acting that distinguishes the individual from other people. Some factors affect and determine personality development, including the person's experiences during early life, the surrounding environment, and the inherited genetic factors. During the early years of life, the child shows a wide variety of behaviors in response to the environment. This response and the way of handling things are unpredictable, changeable according to the results, and occur spontaneously. While the child tries different behaviors to deal with situations, over time, they begin to understand and recognize which behavior or response is permissible and could achieve their desires and resolve the distress and which are terrible choices. Passing through this process over and over will shape the individual personality and their actions with the environment (Zanarini et al. 1998).

A personality disorder is a mental illness disorder characterized by pervasive, inflexible, and maladaptive deviation in the way of thinking, feeling, and acting from the expected individual's culture and causes significant malfunctions and disturbance in the individual relationship that persist over time. It is recognized during late adolescence or early adulthood and continues during life. Usually, the person with personality disorder seeks medical advice due to disagreement with people during work and interpersonal relationships rather than the distress of their thoughts and feelings. These deviations may be ego-syntonic, and they are unaware of and may consider their behavior appropriate as part of their character. These behaviors negatively affect the quality of life and may result in maladaptive defense mechanisms that lead to anxiety and distress.

4 Classification of Personality Disorders

The ten different types of personality disorders are categorized into three clusters that share the same features (Table 1). Each has a distinct pattern of signs and symptoms that differentiate from the others and have its consequences on the individual. We will consider the classification according to the DSM-5.

Table 1 shows clusters A, B, and C of personality disorders according to the DSM-5

	Cluster A	Cluster B	Cluster C
Characteristic behavior	Odd and eccentric	Dramatic, emotional, or erratic	Anxious or fearful
Personality disorder type	Paranoid	Antisocial	Avoidant
	Schizoid	Borderline	Dependent
	Schizotypal	Histrionic	Obsessive-compulsive
		Narcissistic	

4.1 Cluster A Personality Disorder

These types of personality disorders are characterized by odd and eccentric behavior. Individuals with cluster A considered the world overly self-centered; thus, they cannot develop meaningful social relationships. Also, there is a genetic association between this category and schizophrenia, and they are more resistant to treatment. This cluster includes:

1. ***Paranoid Personality Disorder (PPD)***: An individual with PPD considers other people's actions as threatening and suspicious regardless of evidence of a threat, so they have pervasive distrust issues. Also, they may show aggressive behavior without justification because they consider others unfaithful and accuse them of things that don't exist. In addition, PPD is characterized by unforgiving, ruminative, jealous, excessive self-importance, and hostility traits. Individuals with diagnoses of PPD tend to keep observing the environment and other people for signs of a threat to accuse others of planning to get them out, so they are always hypervigilant. Due to this mindset, those individuals have difficulty socializing with people or engaging in romantic relationships because they keep charging their partners to be disloyal. Usually, individuals with PPD cannot locate a problem inside themself because their symptoms are ego-syntonic, so they tend to project their creativity and feelings on other people. That's why the projection defense mechanism is considered to have an essential role in the paranoid process. PPD occurs more commonly in men than women and especially in people from minority cultural statuses. To make the diagnosis of PPD reliable, the DSM-5 proposed two criteria. Criterion A; involve that individual must have a global distrust and suspicion of others' motivations which can be concluded in four of the following features: (1) belief, without sufficient evidence, that other people intentionally mean to deceive and harm them; (2) unjustified doubt about the loyalty and trustworthiness of others; (3) avoid to confide other people due to fearing of unfaithfulness and this information may be used against them; (4) understanding benign remarks or events as meant to be threatening or demeaning; (5) tendency to keep grudges and unforgiving others; (6) recognize attacks on their character which are hidden to others require a vindictive response; and (7) presence of recurrent suspicion and accusing their spouse as being disloyal without justification. At the same time, criterion B suggests that these paranoid symptoms must not be attributed to psychotic episodes associated with schizophrenia, bipolar disorder, a major depressive disorder with psychotic features. Treatment of patients with PPD is quite challenging because they have trust difficulties and fear of being humiliated. However, psychotherapy is worth the try, but first, the therapist has to build a good rapport with the patient. A new promising psychotherapy approach called metacognition-oriented therapy is used to manage those patients by asserting the importance of helping individuals structure narratives of their lives and raising the reflection and metacognition capacities to understand themselves and others, making the patient more aware and improving their interpersonal relationships (Zimmerman and Mattia 1999).

2. ***Schizoid Personality Disorder (SzPD)***: Individuals with this disorder are characterized by social withdrawal, emotionally cold, carelessness about other people's opinions, avoidance of intimate sexual relationships with their partner, looking to the world as an observer, not as a part of it, and avoid social situations or occupations that require them to be engaged in teamwork. Due to this disconnection and fear from the social world, the individual with SzPD may fall back into one's internal fantasy or isolate themselves as a way of defense. Prevalence of SzPD is low in the general population and tends to range between 1 and 5%, according to the American Psychiatric Association (2013). There is evidence of a link between genetic susceptibility to SzPD and schizophrenia. Features of isolation, lack of relationship with peers, poor school performance, and annoyance by others during childhood may predict the development of SzPD later in life. Diagnosis of SzPD is applied through DSM-5 criteria, which involve Criterion A, defined by detachment from a social and familial relationship with a restricted range of emotional expressions in an interpersonal setting that starts at or after the age of 18. While criterion B proposed the mentioned features must not be associated with the course of schizophrenia, bipolar disorder, a depressive disorder with psychotic symptoms, or autism spectrum disease. As is the case with most personality disorders, individuals diagnosed with SzPD rarely look for treatment because their symptoms are ego-syntonic and don't interfere with their self-esteem. In fact, there is no specific treatment approved for the management of SzPD. However, like other personality disorders, using psychotherapy may improve the symptoms and help individuals with SzPD develop social skills. In psychotherapy, the therapist tries to make the patient observe their maladaptive behavior then encourage using new skills to counteract this behavior. Group therapy may be used as adjunctive treatment to help patients interact with others to reduce their fear and identify their positive and negative emotions. Because the prevalence of anxiety and depression is higher in people with SzPD, pharmacological treatment may be used in these cases (Leichsenring et al. 2011).
3. ***Schizotypal Personality Disorder (SPD)***: Individuals with this disorder have an eccentric appearance, unconventional beliefs, distorted or magical thinking, and interpersonal awkwardness. Also, people with this mental disease complain of social difficulties and lack of relationships outside their family due to feelings of pronounced discomfort. Due to a combination of social deficits and cognitive deterioration (involves primary working memory, verbal learning, context processing, and sustained attention), subjects with SPD suffer significantly poor functioning and reduced quality of life in society. Although this disorder is not widely investigated, there is some evidence suggesting low prevalence, and it ranges from less than 1% to 4%, with a slightly higher rate in men than women. During childhood, symptoms of SPD manifest by excessive social anxiety, sensitivity, and eccentric thoughts and speech. SPD is also considered a prototype form of schizophrenia spectrum disorders. To deal with unconscious conflicts, individuals with SPD use immature defense mechanisms predominated by emerging projection and passive aggression defenses to manage the situation.

Also, the presence of autistic fantasy, displacement, and passive aggression defense mechanisms may predict the development of SPD later on. It may be challenging to differentiate between SPD, PPD, and BPD in a clinical setting due to the overshadowing of the symptoms. However, it may be helpful to do that via using the diagnostic criteria of each of them. To make SPD diagnosis clear, two criteria proposed by DSM-5 must be present. Criterion A characterized by pervasive social, interpersonal, and perceptual deficits. At the same time, criterion B mentioned that these features must not correspond to the course of schizophrenia, bipolar disorder or depressive disorder with psychotic symptoms, or another psychotic disorder. Treatment of subjects with SPD involves mainly psychotherapy. Although there is limited data on psychopharmacological agents, it could be used as an adjunct to psychotherapy. For example, psychotic-like symptoms (odd belief, unusual perceptual experiences, idea of reference, magical thinking, and suspiciousness) can be managed using small doses of second-generation antipsychotic drugs (e.g., risperidone or olanzapine—also decrease mood symptoms). In the presence of concomitant depression or mood symptoms, some data suggest using antidepressant agents to mitigate them. Psychotherapy is the same for other personality disorders that involve establishing an alliance as the first step because patients with this disorder usually have no desire to make a relationship. More suspicion could be a barrier to successful therapy. There are no trials on the effectiveness of different types of psychotherapy, but some experts recommend using supportive, supportive-expressive, or exploratory-oriented approaches. Also, some reports demonstrate the effectiveness flexibility of metacognitive interpersonal therapy and metacognitive reflection and insight therapy. The use of cognitive-behavioral therapy is still helpful. It allows the individual with SPD to focus on building social skills, relieving mood symptoms, being more aware of one's behavior, and maintaining core beliefs.

4.2 Cluster B Personality Disorders

This personality disorder class has dramatic, emotional, or erratic behavioral features. Individuals with disorders in this group have predominantly manifested by lack of empathy, and they tend to involve themselves in impulsive, promiscuous, and illegal experiences. Also, the disorders in this cluster have a genetic association with mood disorders and substance use. This cluster includes:

1. ***Antisocial Personality Disorder (ASPD)***: Characterized by a chronic pattern of violation of other people's rights with lack of remorse and tendency to criminality, responsibility, impulsivity, hostility, and manipulation of others for personal interest. Individuals with ASPD usually fail to develop stable interpersonal relationships because they are dominated by aggressiveness and violate other people's rules. In addition, they don't obey the law, so they may have a history of

legal problems and fail to sustain consistent employment. ASPD originates during childhood or early adolescence due to physical, emotional, and sexual maltreatment and continues into adulthood. The criteria of DSM-5 can confirm the diagnosis of ASPD. Criterion A: consisted of an enduring pattern of disregard for and violation of other rights that begin at 15 years of age. The effectiveness of psychotherapy is questionable since the disorder tends to diminish with time. Some studies conduct the only treatment is a passage of time. Some literature suggests cognitive-behavioral therapy for ASPD; however, due to the patient's impulsivity and aggressiveness, the therapist has difficulty establishing an alliance, so it has limited benefit. Treating conductive disorder during childhood may be the best approach to minimize ASPD symptoms during adulthood. Psychopharmacological agents (e.g., antidepressants, antipsychotics, and mood stabilizers) could be used only to control symptoms of aggression and impulsivity and manage the concomitant disorder. An approach called mentalization-based therapy (involving thinking about self mental state and mental state of others) has emerged in recent years that appears to be effective in minimizing aggressiveness. However, more data is needed to confirm these findings.

2. ***Borderline Personality Disorder (BPD)***: An enduring pattern of instability prominent in mood, interpersonal relationships, and self-concept perspectives, and associated with the feeling of emptiness, fear of abandonment, engaging in impulsive behaviors, and trying to get others attention via self-mutilation and recurrent suicidal attempt. Individuals with BPD tend to recognize the world into the extremes of either "all good" or "all bad." The literature revealed that individuals with BPD exhibit a higher rate of using immature defense mechanisms that include acting out, emotional hypochondriasis, and undoing, so they may use them as screening tools in a clinical setting to recognize patients with BPD. Most recent studies demonstrate that using splitting, projective identification, and affiliation defenses associate with a higher rate of suicide among people with BPD. The DSM-5 defines BPD as a pervasive pattern of instability of interpersonal relationships, self-image, affects and marked impulsivity that starts early adulthood and exists in various contexts. Many modalities can be used and proved to be successful such as dialectical behavioral therapy (DBT), transference-focused psychotherapy (TFP), and mentalization-based therapy (MBT). Sometimes medications could be used as an adjunct to psychotherapy to control symptoms of BPD, such as small doses of second-generation antipsychotics for managing perceptual symptoms and selective serotonin reuptake inhibitors to reduce impulsivity and aggression behaviors (Lewis and Grenyer 2009; Lieb et al. 2004; Vaillant 1992).
3. ***Narcissistic Personality Disorder (NPD)***: Characterized by an enduring pattern of grandiosity that causes impairment in interpersonal relationships, a requirement of excessive attention and admiration, exaggerated sense of self-centeredness, and hypersensitivity to criticism. Individuals with this disorder also have a significant psychosocial disability and low quality of life. Although NPD is associated with many other mental health disorders, it can coexist with substance use disorders and other personality disorders like ASPD, BPD, HPD,

SPD. Some studies classified NPD into two subtypes: grandiose NPD and vulnerability NPD. Grandiose NPD is characterized by aggression, excessive grandiosity, boldness, arrogance, and self-inflated image. DSM-5 defines NPD as a pattern of grandiosity (in fantasy or acting), constantly seeking admiration, and lack of empathy, which appears in early adulthood and present in different situations. For NPD treatment, no data establishes a single model to be highly effective and efficient, but a nice try of psychotherapy may be beneficial. Some experts suggest the same psychotherapeutic approach used for BPD, but it is not confirmed yet and needs more investigations. Severe symptoms of NPD could be mitigated by mood stabilizer agents, antidepressant, and antipsychotic medications (Vaillant 1977; Costa 2016, 2017).

4. ***Histrionic Personality Disorder (HPD)***: Characterized by patterns of attention-seeking behaviors associated with dramatic speech and exaggerated theatrical, emotional expression, and could be easily provocative sexually. People with HPD always try to be the center of attention by their physical actions because they may feel disregarded or depressed if they are not. They also tend to appear fascinating, tempting, lively, enthusiastic, and manipulative to achieve their desires. To make the diagnosis of HPD feasible, the DSM-5 defines HPD as a pervasive pattern of excessive emotional dysregulation and attention-seeking that starts in early adulthood and persists over time in a variety of contexts, then proposed eight criteria, five or more of them required to ensure the diagnosis. The subjects with NPD unconsciously being dramatic as a way of dealing with their internal conflicts and anxiety. For example, they adopt opinions of people they admire as a fact without considering or thinking about themselves. Unfortunately, there is no significant data about the best effective treatment for NPD. However, psychotherapy may be the best option available that the patient can benefit from in the long term. Supportive psychotherapy may aid people with NPD to enhance their emotional control, increase their self-esteem, and better perceive and cope with the environment. Cognitive-behavioral therapy may also be used in conjunction with psychodynamic therapy to increase the effectiveness of the treatment. The goal of psychodynamic psychotherapy is to reduce the unconscious internal conflicts as an attempt to make the patients better understand themselves and recognize their behaviors by learning to exchange dramatic speech into more adaptive behaviors. Also, the same pharmacological agents that are used in NPD could be used in severe cases to relieve the symptomatic dysregulations.

4.3 Cluster C Personality Disorders

Anxious and fearful thoughts and behavior characterize this category of personality disorders. It occurs due to a combination of genetic and environmental etiologies. Individuals who have a disorder from this cluster suffer from poor relationships and impairment in occupational functioning. These disorders are also genetically associated with anxiety disorders. This cluster consists of three disorders, include:

1. ***Avoidant Personality Disorder (AVPD)***: Characterized by excessive social discomfort, hypersensitivity to rejection and criticism, sense of inadequacy, avoidance of social activities and interpersonal relationships, fears of embarrassment that prevent the individual from taking a risk or being involved in new activities. Individuals with AVPD have prominent impairment in mental, social, and somatic aspects that lead to more difficulty in daily life than other personality disorders. It is proven that this disorder has a developmental deficit during infancy or early childhood that makes the individual appear shy and prefer isolation. Also, exclusive studies link AVPD and social anxiety disorder (SAD). Some reports suggest that about 40% of AVPD patients have SAD, while others have higher rates. However, some literature considers SAD a subtype of AVPD, while others regard it as a distinct entity with overlapping features. DSM-5 defines AVPD as a consistent pattern of social inhibition, feeling of inadequacy, and being excessively sensitive to negative evaluation that begins at early adulthood and continues in different life aspects. Although few studies are conducted to investigate the best approach to treat people with AVPD, certain modalities used to treat SAD effectively reduce the burden of the AVPD and alleviate the symptoms. Cognitive-behavioral therapy (CBT) is one of these modalities that targets negative thoughts and how these thoughts can badly affect the behaviors. Then, try to figure out new skills to correct these behaviors through specific techniques (e.g., graded exposure, advantages and disadvantages of avoidance, and behavioral experiment). Schema therapy is another modality that uses a combination of the already existing techniques and theoretical bases of CBT, psychoanalytic therapy, attachment theory, and Gestalt theory to help the patients develop more sense of self-esteem and learn their basic needs and how to achieve them adaptively. Lastly, the studies show that psychodynamic therapy is generally adequate for cluster C personality disorders and not exclusively for AVPD. Although some experts recommend using some pharmacological agents that are used to manage SAD-like selective serotonin reuptake inhibitors, serotonin-norepinephrine reuptake inhibitors, and antidepressants, there are no medications approved by the authorities to manage AVPD (Vaillant 2016; Walker and Lim 2010).
2. ***Dependent Personality Disorder (DPD)*****:** Characterized by an excessive need for support, low self-confidence, dependence on others to direct them, fear of rejection, submissiveness, and passive behavior. Individuals with DPD excessively rely on other people to take care of them and fulfill their physical and psychological needs as they perceive themselves as weak and ineffectual. They are also stuck in abusive relationships and feel disappointed and hopeless when broken up in a relationship. The American Psychiatric Association estimates the prevalence rate of DPD to be about 0.49%, while other studies estimate it to range from 1 to 5% in the general population with more affected women than men. People with DPD are more likely to ask for treatment and guidance than other personality disorders because they always try to find someone to care for them.

Also, because they are hypersensitive to interpersonal cues, enthusiastic to please, and capable of forming good alliances, these features may aid and ease the treatment to achieve better outcomes. There is no specific treatment for DPD. The general treatment for any personality disorder may be beneficial, such as psychotherapy approaches to help individuals with DPD improve their self-esteem and confidence, thus being independent of others and building more positive relationships. At the same time, psychodynamic approaches aim to reveal unconscious conflicts and make them accessible to foster insights. Lastly, behavioral therapy aims to identify and recognize the specific dependent behaviors then replace them with more adaptive and autonomous acts. One problem with treating DPD patients is that they may have difficulty terminating the treatment due to fear of abandonment, so the therapist should be careful when dealing with those patients (Patrick and Diamond 2017; Mulay and Cain 2020; Fisher and Hany 2021; French and Shrestha 2020).

3. ***Obsessive-Compulsive Personality Disorder (OCPD)***: Characterized by chronic maladaptive behaviors that consist of one's own beliefs and attitudes, preoccupation with orderliness, seek perfectionism. Individuals with OCPD are conscientious and have high standards, so they are unsatisfied with their achievements. Although they are reliable to do many tasks and responsibilities, it may be challenging to adapt to changes because they are inflexible. They also consider every detail and be more cautious, which makes it sometimes uncomfortable and challenging for them to decide and function properly. The studies estimate that OCPD is the most common personality disorder, with a prevalence rate ranging from 7% to 8% in the general population. In addition, OCPD is highly associated with the co-occurrence of other personality disorders like PPD, SzPD, SPD, and ASPD. Usually, symptoms of people with OCPD are ego-syntonic, so they don't regard themselves as having problems and thus don't seek any treatment. In contrast, other studies suggest contradictory findings that people with OCPD seek treatment because of their suffering and impairment of functions. Like other personality disorders, there is a lack of studies on OCPD treatment exclusively. However, most data available from co-occurring disorders have shown to be effective to some degree. For example, studies conducted on OCPD that coexist with depression and anxiety using cognitive therapy (CT) have decreased the severity of OCPD and other disorders. Studies that test cognitive-behavioral therapy (CBT) in people with OCPD and mood disorders have also shown improvement in both disorders with decreased levels of distress. Other modalities such as dialectical behavioral therapy (DBT), metacognitive interpersonal therapy (MIT), schema therapy, and supportive-expressive, dynamic psychotherapy also could be used to manage individuals with OCPD. Still, more randomized studies need to be conducted to declare the effectiveness of each of them.

Multiple Choice Questions

1. A 32-year-old female presented to the psychiatric clinic for assessment due to a third-time suicidal attempt after breaking up with her romantic partner 2 months ago. Since that time, she has engaged in unsafe sexual behaviors and drinks excessively. Also, she suffers from depression and intense feelings of emptiness. The most likely diagnosis is:
   (a) Paranoid personality disorder
   (b) Narcissistic personality disorder
   (c) Borderline personality disorder
   (d) Histrionic personality disorder
   (e) Dependent personality disorder

2. A person diagnosed with a schizotypal personality disorder. The person has all of the following features, except:
   (a) Eccentric appearance and magical thinking
   (b) Use passive aggression and autistic fantasy as main defense mechanisms to deal with stressful situations
   (c) Suspicious and paranoid ideation
   (d) The main treatment is psychotherapy
   (e) Has no desire for or enjoyment in interpersonal relationships

3. Which of the following features are present in a patient with the obsessive-personality disorder:
   (a) Difficulty in starting projects on his own due to lack of self-confidence
   (b) Socially restricted
   (c) Inability to complete tasks sometimes due to seeking perfection
   (d) Uncomfortable when not being the center of attention
   (e) The belief that the person is unique and knowledgeable

4. All of the followings are features of antisocial personality disorder, except:
   (a) Violation of other people's rights and lack of remorse
   (b) Believe without sufficient evidence that other people want to hurt them
   (c) Repeated lying and try to use people to get personal benefit
   (d) Associated with substance abuse, bipolar disorder, and attention deficit hyperactivity disorder
   (e) The best treatment option to reduce the symptoms is to treat conductive disorder during childhood

5. A 12-year-old child was presented to the psychiatric unit by his/her parents because they were worried about him after their divorce. He/she does not talk to them instead. He/she fights in the school with his/her friends and doesn't obey his/her teacher's instructions. Which defense mechanism he/she exhibits by his behavior:
   (a) Acting out
   (b) Passive aggression
   (c) Splitting
   (d) Reaction formation
   (e) Displacement

6. A mother brought her 5-year-old daughter to the emergency department due to abdominal pain and vomiting. After doing the required investigations, the doctor diagnosed her with diabetes mellitus type I. The mother refused to accept this diagnosis and requested another doctor's opinion because she has an older daughter with the same disease and cannot deal with both of them. What is the defense mechanism the mother used:
   (a) Displacement
   (b) Schizoid fantasy
   (c) Denial
   (d) Idealization
   (e) Intellectualization
7. A 20-year-old girl was exposed to rape. When the police officer asked her about the event, she described it in detail without any emotional act. What defense mechanism she used:
   (a) Dissociation
   (b) Isolation of affect
   (c) Repression
   (d) Spitting
   (e) Projection
8. A wife who divorced her husband is looking for online divorce data to show her friends the divorce rate to avoid the feeling of sadness and discomfort of the situation. What defense mechanism represents the wife's behavior:
   (a) Pass aggression
   (b) Projection
   (c) Reaction formation
   (d) Intellectualization
   (e) Sublimation
9. All of the following are features of mature defense mechanism, except:
   (a) Used by a healthy individual to handle uncomfortable daily events to relieve anxiety
   (b) May indicate more emotionally intelligent individual and reasonable self-satisfaction
   (c) Used consciously by the individual to control life events
   (d) The reality is kept itself without distortion
   (e) Associated with schizophrenia and borderline personality disorder
10. A wife who despises her husband's daughter instead cares about her and is very friendly and lovely with her. Which defense mechanism represents the wife's behavior:
   (a) Reaction formation
   (b) Sublimation
   (c) Altruism
   (d) Suppression
   (e) Isolation of affect

Answers

1. (c)
2. (e)
3. (c)
4. (b)
5. (a)
6. (c)
7. (d)
8. (d)
9. (e)
10. (a)

References

American Psychiatric Association. Diagnostic and statistical manual of mental disorders. 5th ed. Washington, DC: Author; 2013.

Caligor E, Levy KN, Yeomans FE. Narcissistic personality disorder: diagnostic and clinical challenges. Am J Psychiatr. 2015;172(5):415–22.

Clarkin JF, Caligor E, Sowislo J. TFP extended: development and recent advances. Psychodyn Psychiatry. 2021;49(2):188–214.

Costa RM. Conversion (defense mechanism). In: Encyclopedia of personality and individual differences. Cham: Springer International Publishing; 2016. p. 1–4.

Costa RM. Undoing (defense mechanism). In: Encyclopedia of personality and individual differences. Cham: Springer International Publishing; 2017. p. 1–2.

Esterberg ML, Goulding SM, Walker EF. Cluster a personality disorders: schizotypal, schizoid and paranoid personality disorders in childhood and adolescence. J Psychopathol Behav Assess. 2010;32(4):515–28.

Fisher KA, Hany M. Antisocial personality disorder. In: StatPearls. Treasure Island: StatPearls Publishing; 2021.

French JH, Shrestha S. Histrionic personality disorder. In: StatPearls. Treasure Island: StatPearls Publishing; 2020.

Frick PJ. Early identification and treatment of antisocial behavior. Pediatr Clin. 2016;63(5):861–71.

Iacovino JM, Jackson JJ, Oltmanns TF. The relative impact of socioeconomic status and childhood trauma on black-white differences in paranoid personality disorder symptoms. J Abnorm Psychol. 2014;123(1):225.

Lee RJ. Mistrustful and misunderstood: a review of paranoid personality disorder. Curr Behav Neurosci Rep. 2017;4(2):151–65.

Lee YJ, Keum MS, Kim HG, Cheon EJ, Cho YC, Koo BH. Defense mechanisms and psychological characteristics according to suicide attempts in patients with borderline personality disorder. Psychiatry Investig. 2020;17(8):840.

Leichsenring F, Leibing E, Kruse J, New AS, Leweke F. Borderline personality disorder. Lancet. 2011;377(9759):74–84.

Lewis KL, Grenyer BF. Borderline personality or complex post-traumatic stress disorder? An update on the controversy. Harv Rev Psychiatry. 2009;17(5):322–8.

Lieb K, Zanarini MC, Schmahl C, Linehan MM, Bohus M. Borderline personality disorder. Lancet. 2004;364(9432):453–61.

McGauley G, Yakeley J, Williams A, Bateman A. Attachment, mentalization and antisocial personality disorder: the possible contribution of mentalization-based treatment. Eur J Psychother Counsell. 2011;13(4):371–93.
Morrison J. Histrionic personality disorder in women with somatization disorder. Psychosomatics. 1989;30(4):433–7.
Mulay AL, Cain NM. Antisocial personality disorder. In: Encyclopedia of personality and individual differences. Cham: Springer International Publishing; 2020. p. 206–15.
Nestadt G, Romanoski AJ, Chahal R, Merchant A, Folstein MF, Gruenberg EM, McHugh PR. An epidemiological study of histrionic personality disorder. Psychol Med. 1990;20(2):413–22.
Novais F, Araújo AM, Godinho P. Historical roots of histrionic personality disorder. Front Psychol. 2015;25(6):1463.
Patrick E, Diamond E. Id. In: Encyclopedia of personality and individual differences. Cham: Springer International Publishing; 2017. p. 1–2.
Rnic K, Dozois DJ, Martin RA. Cognitive distortions, humor styles, and depression. Eur J Psychol. 2016;12(3):348.
Ronningstam E. Narcissistic personality disorder in DSM-V—in support of retaining a significant diagnosis. J Personal Disord. 2011;25(2):248–59.
Salvatore G, Russo B, Russo M, Popolo R, Dimaggio G. Metacognition-oriented therapy for psychosis: the case of a woman with delusional disorder and paranoid personality disorder. J Psychother Integr. 2012;22(4):314.
Santana MR, Zatti C, Spader ML, Malgarim BG, Salle E, Piltcher R, Ceresér KM, Bastos AG, Freitas LH. Acute stress disorder and defense mechanisms: a study of physical trauma patients admitted to an emergency hospital. Trends Psychiatry Psychother. 2017;4(39):247–56.
Shea MT, Stout RL, Yen S, Pagano ME, Skodol AE, Morey LC, Gunderson JG, McGlashan TH, Grilo CM, Sanislow CA, Bender DS. Associations in the course of personality disorders and Axis I disorders over time. J Abnorm Psychol. 2004;113(4):499.
Sulz S, Hysteria I. Histrionic personality disorder. A psychotherapeutic challenge. Der Nervenarzt. 2010;81(7):879–87.
Thomas AG, Jonason PK, Blackburn JD, Kennair LE, Lowe R, Malouff J, Stewart-Williams S, Sulikowski D, Li NP. Mate preference priorities in the east and west: a cross-cultural test of the mate preference priority model. J Pers. 2020;88(3):606–20.
Tredget JE. The aetiology, presentation and treatment of personality disorders. J Psychiatr Ment Health Nurs. 2001;8(4):347–56.
Vaillant G. Adaptation to life. Cambridge: Harvard University Press; 1977. 1977 [cited 2021 Sep 9]. https://opus4.kobv.de/opus4-Fromm/frontdoor/index/index/docId/28960
Vaillant GE. Ego mechanisms of defense: a guide for clinicians and researchers. Arlington: American Psychiatric Association Publishing; 1992.
Vaillant GE. Defense mechanisms. In: Encyclopedia of personality and individual differences. Cham: Springer International Publishing; 2016. p. 1–10.
Vos MS, De Haes JC. Denial in cancer patients, an explorative review. Psychooncology. 2007;16(1):12–25.
Walker T, Lim BK. Defense/coping mechanism. In: Encyclopedia of cross-cultural school psychology. Boston: Springer US; 2010. p. 361–3.
Yazdanshenas Ghazwin M, Tavakoli SA, Latifi S, Saberi H, Derakhshanrad N, Yekaninejad MS, Sadeghi M, Emami Razavi SH, Norouzi Javidan A, Ghodsi SM. Psychological defense mechanisms among individuals with SCI with adjustment disorder. J Spinal Cord Med. 2017;40(5):538–47.
Zanarini MC, Frankenburg FR, Dubo ED, Sickel AE, Trikha A, Levin A, Reynolds V. Axis I comorbidity of borderline personality disorder. Am J Psychiatr. 1998;155(12):1733–9.
Zanarini MC, Weingeroff JL, Frankenburg FR. Defense mechanisms associated with borderline personality disorder. J Personal Disord. 2009;23(2):113–21.

Zanarini MC, Frankenburg FR, Fitzmaurice G. Defense mechanisms reported by patients with borderline personality disorder and axis II comparison subjects over 16 years of prospective follow-up: description and prediction of recovery. Am J Psychiatr. 2013;170(1):111–20.

Zimmerman M, Mattia JI. Axis I diagnostic comorbidity and borderline personality disorder. Compr Psychiatry. 1999;40(4):245–52.

Drugs and Consciousness

Mustafa Hayder Kadhim

1 CNS Stimulants

They stimulate the CNS and. These drugs include caffeine which can be found in coffee and chocolate (Fig. 1). It mimics adenosine (a natural brain chemical) (Weil 1998).

Many students who consume many cups of coffee experience suffering from headaches due to the sudden decrease in the levels of caffeine in their brains (Fig. 2).

Nicotine which is found in cigarettes is a stimulant. Smokers experience an addictive effect on nicotine, and in the same way, when they suddenly stop smoking develop a withdrawal effect. Nicotine is an activator, activates neural mechanisms, and shows similarity to cocaine.

1.1 Amphetamine

Dexedrine and Benzedrine are strong stimulants. They stimulate the central nervous system and lead to energy and alertness, talkativeness, insomnia, and wakefulness. They increase concentration and reduce fatigue (Koob and Volkow 2010).

Chemical structure: $C_9H_{13}N$

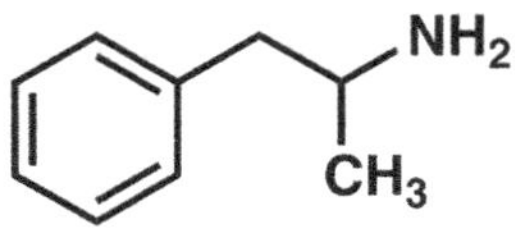

M. H. Kadhim (✉)
University of Kufa, College of Medicine, Najaf, Iraq

H. T. Hashim, A. Alexiou (eds.), *The Psychology of Consciousness: Theory and Practice*, https://doi.org/10.1007/978-3-030-90692-4_6

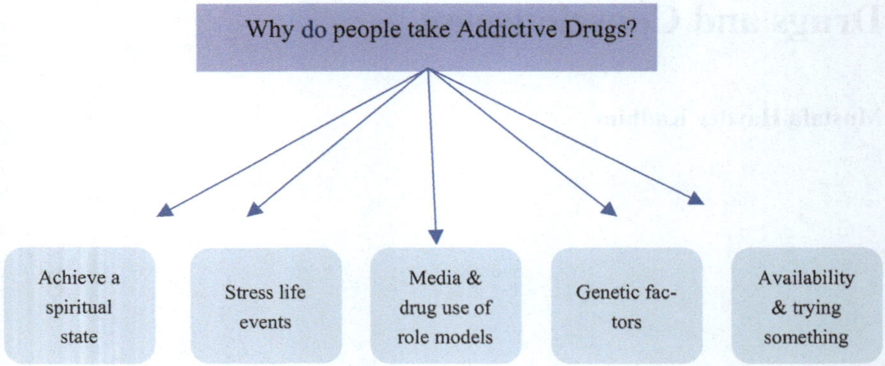

Fig. 1 Addiction

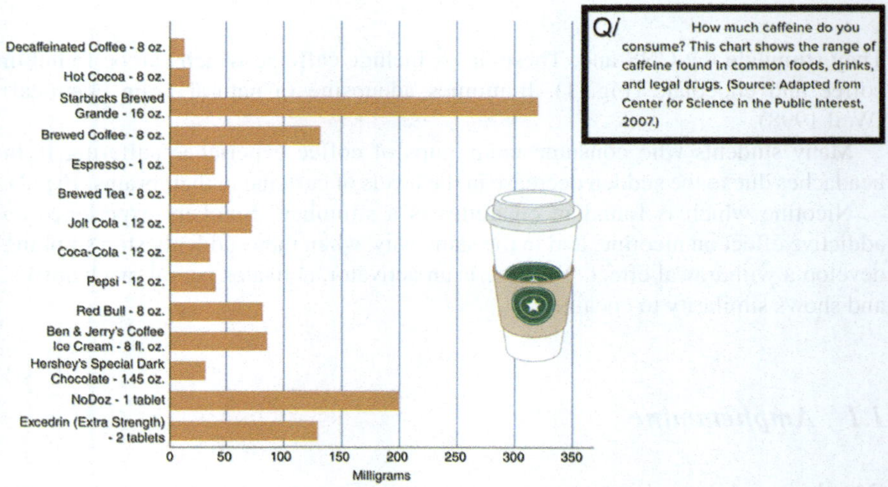

Fig. 2 The level of caffeine in the common drugs and foods

Mechanism of action: Psychostimulant. It acts centrally and causes releasing of DA, NE, 5HT.

Clinical uses: Narcolepsy, attention deficits, hyperactivity disorder, obesity, chronic pain, depression.

Many randomized clinical trials and research have proven its effectiveness and safety in treating ADHD and improving cognitive and physical performance.

Methamphetamine is a white, crystalline drug; it is one of the most dangerous street drugs used nowadays. Moreover, it is highly addictive and cheap.

Bath salts are also an amphetamine-like stimulant. It can produce euphoria and a rise in sociability and sex drive, but the side effects can be severe, including paranoia and agitation (Koob and Volkow 2010).

1.2 Cocaine

Cocaine is another strong central nervous system stimulant. Fortunately, it is not used anymore but still forms a major concern because of its illegal usage. It can be smoked, inhaled, or injected into the veins. It acts immediately and absorbs quickly.

Cocaine produces profound psychological well-being, increased confidence, decreased appetite, anxiety, irritability, and insomnia.

Sudden withdrawal signs and symptoms include suicidal thoughts, which is severe and agitated motor activity, and irritability.

Cocaine is considered an addictive drug because of its effect on the reward pathway in the CNS. It also increases respiratory disease, MI, stroke (especially in those who smoke), and blood infections (Koob and Volkow 2010).

Chemical structure: $C_{17}H_{21}NO_4$

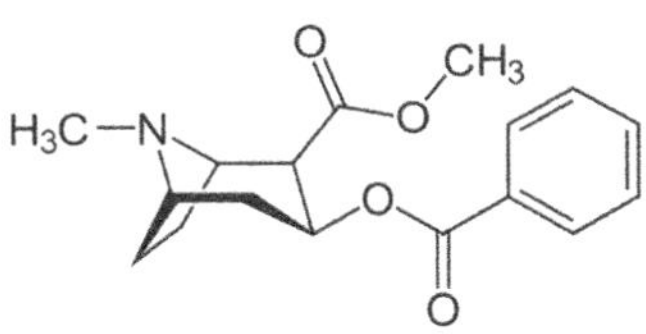

Mechanism of action: NE, DA, 5HT reuptake inhibitor.

Clinical uses: Local anesthetic.

The symptoms and signs of cocaine toxicity are not just limited to certain limits. At first, psychiatric complaints (depression potentiated by dysphoria, CNS excitation, and paranoia), seizures, and even chest pain (Lindesmith 2017).

2 CNS Depressants

2.1 Alcohol

It is a depressant, and it is used extensively worldwide and among almost all people (Feldman 2011; Su et al. 1997; Grant 1997).

Its effects are comprehensive (Fig. 3). Addiction comes from the idea of relieving tension and making the person feel happy and disconnected from the world (Feldman 2011; Anthenelli and Schuckit 1993).

2.2 *Barbiturates*

"Barbiturates include drugs such as phenobarbital, frequently prescribed by physicians to induce sleep or reduce stress; barbiturates produce a sense of relaxation" (Wagner and Anthony 2002). However, they produce altered thinking, faulty judgment, and sluggishness at larger doses. In addition, it is deadly when combined with alcohol.

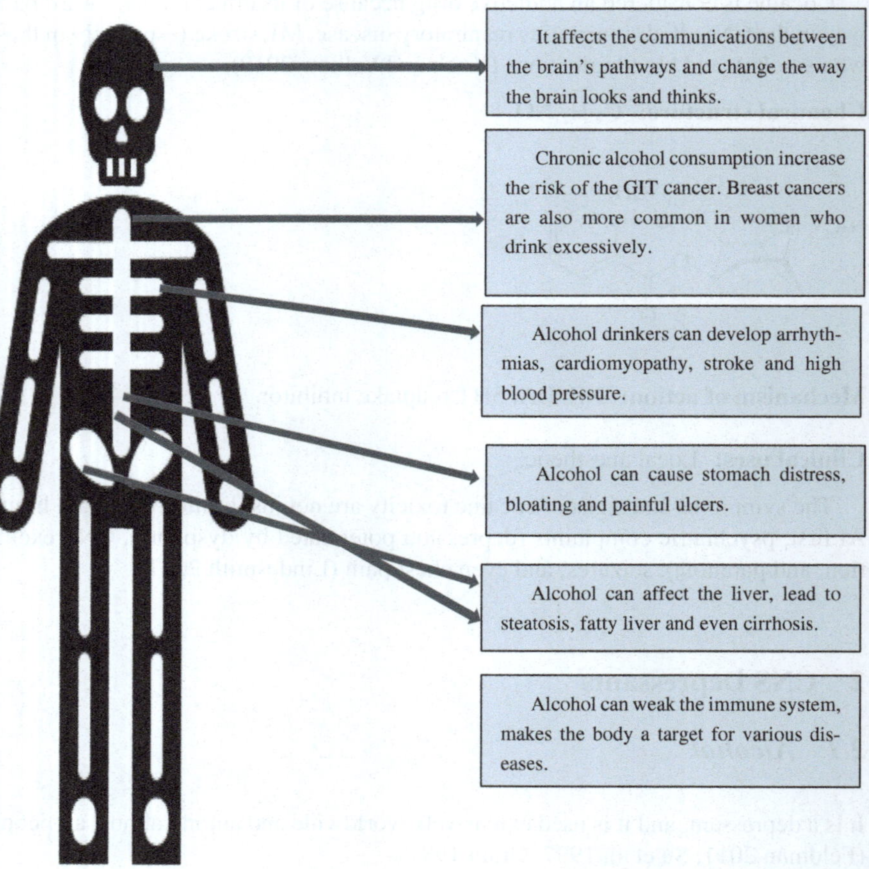

Fig. 3 Alcohol effects

Chemical structure:

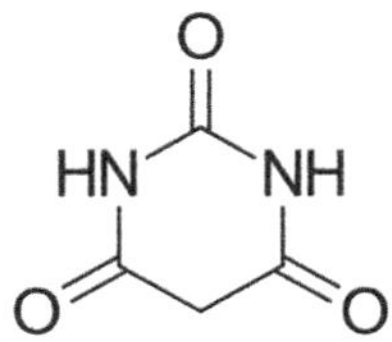

Barbiturates work as a GABA agonist (GABA is an inhibitory neurotransmitter). It prolongs GABA activity and increases its duration (Clark et al. 2011; Cami and Farré 2003).

3 Narcotics

They are a wide range of drugs that are used to relieve stress and tension and to change the mood, and it causes addiction very widely (Wang 2017; Clark et al. 2011; Beaudoin et al. 2016).

3.1 Morphine

Chemical structure: $C_{17}H_{19}NO_3$

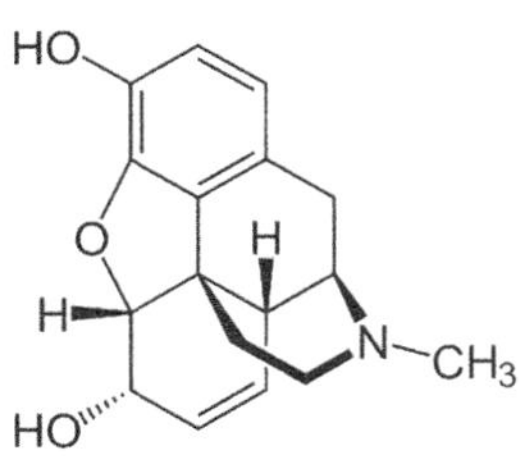

Medical uses: Morphine is so vital as an analgesic in acute MI. It is considered lifesaving (relieves pain without affecting the consciousness). Patients given morphine are still aware of pain, but the sensation is not unpleasant (Fig. 4).

Both morphine and codeine can prevent cough because they have antitussive effects (Grissinger 2011; Prus 2018).

Morphine has significant effects on endocrine properties in the body (Yilmaz et al. 1999).

The conjugates that have been mentioned above are excreted mainly in the urine, with small quantities appearing in bile.

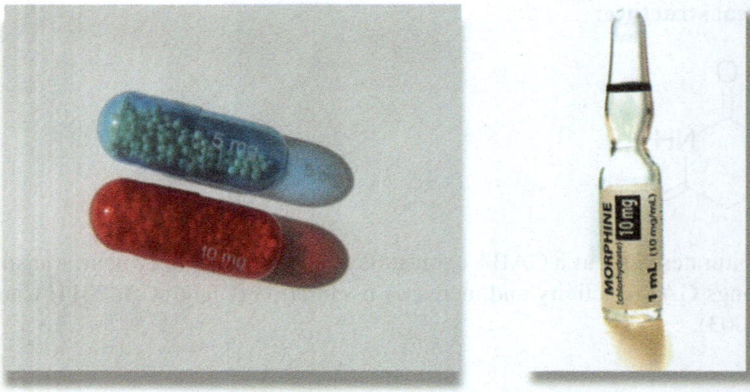

Fig. 4 Morphine distribution

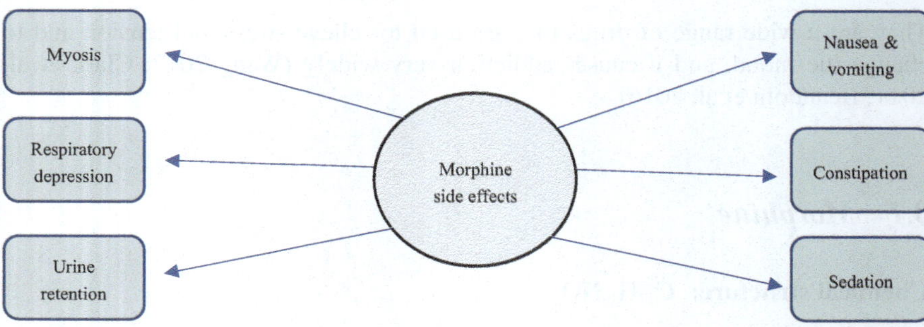

Fig. 5 Morphine side effects

Adverse effects (Fig. 5) with most μ agonists, severe respiratory depression is one of the biggest concerns you need to be vigilant for (Schiller et al. 2021; Imani et al. 2014).

3.2 *Codeine*

It is a natural opioid that can be used in analgesics, but it is weak. It is a selective agonist for mu receptors of opioids.

Chemical structure: $C_{18}H_{21}NO_3$

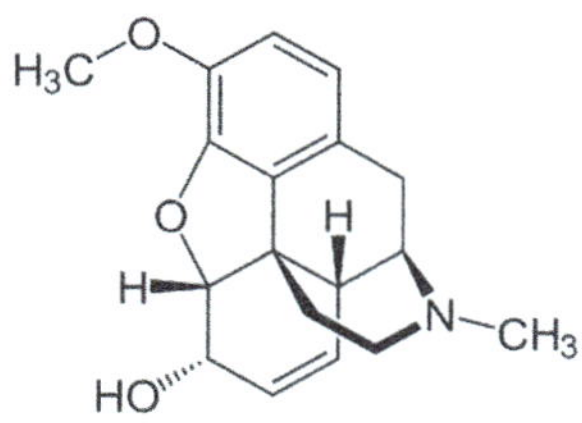

Side effects: As with other opioids, it causes nausea, vomiting, constipation, drowsiness, euphoria and dysphoria, meiosis, urinary retention.

Also, tolerance may develop in chronic users, but it is slowly developed to constipation and meiosis.

3.3 *Fentanyl*

It is more potent than morphine in analgesic effects and used in anesthesia (Fig. 6) (National Center for Biotechnology Information 2021).

Chemical structure: $C_{22}H_{28}N_2O$

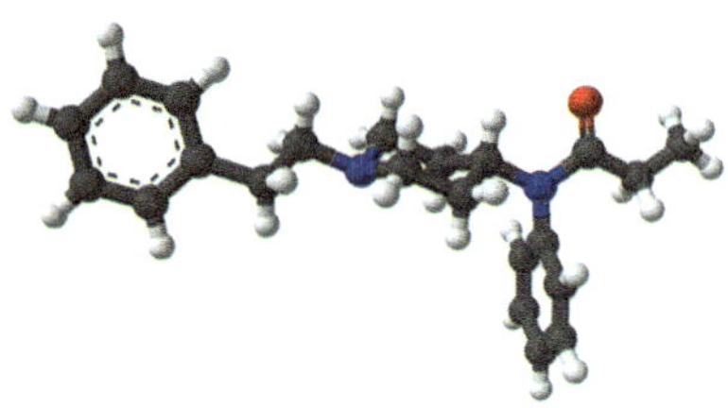

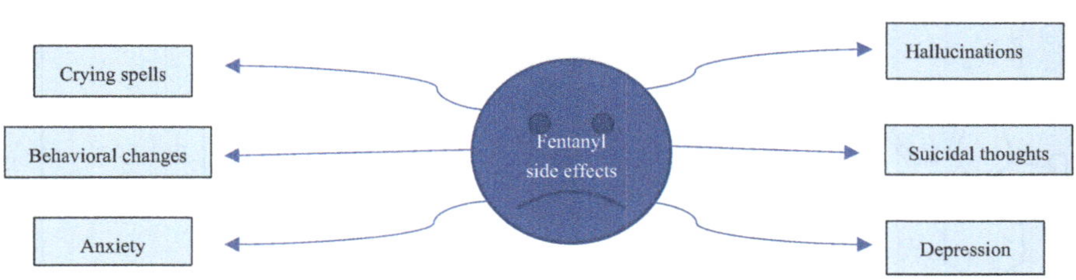

Fig. 6 Fentanyl side effects

3.4 Methadone

Methadone is a synthetic opioid administered orally (Fredheim et al. 2008).

Chemical structure: $C_{21}H_{27}NO$

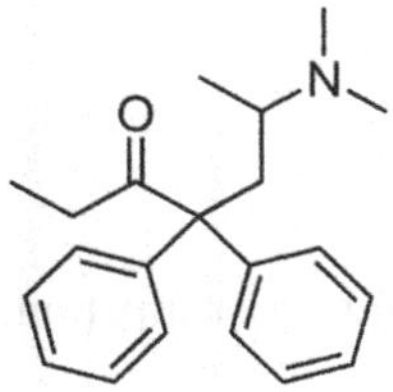

Individuals with a history of prolonged QT should be used with high caution because it can further prolong the QT interval and lead to serious life events (Ahmed et al. 2016).

3.5 Meperidine

It is a synthetic opioid with lower potency, and it is considered a weak opioid. It has no relation to morphine. Meperidine is a highly lipophilic drug and has anticholinergic properties, increasing the incidence of delusions and delirium compared to other opioids (Baldo 2018).

Pharmacokinetics: The duration of action of meperidine is shorter compared to morphine and other opioids.

Drug interactions: Meperidine should be used with caution in concomitant or recent administration of MAOIs selective serotonin reuptake inhibitors (SSRIs) because it may lead to serotonin syndrome, which leads to shivering, diarrhea, seizures, rigidity, and fever.

Chemical structure: $C_{15}H_{21}NO_2$

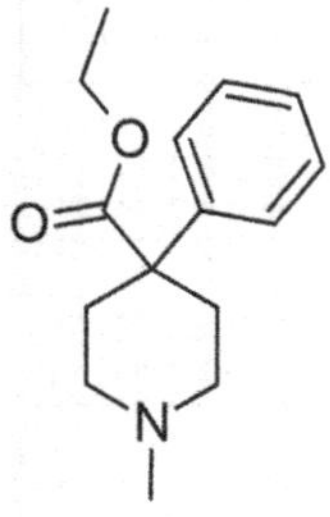

3.6 *Oxycodone*

Oxycodone is a semisynthetic derivative of morphine (National Center for Biotechnology Information 2021).

Toxicity: Abuse or having toxic doses can lead to death (Söderberg Löfdal et al. 2013).

Chemical structure: $C_{18}H_{21}NO_4$

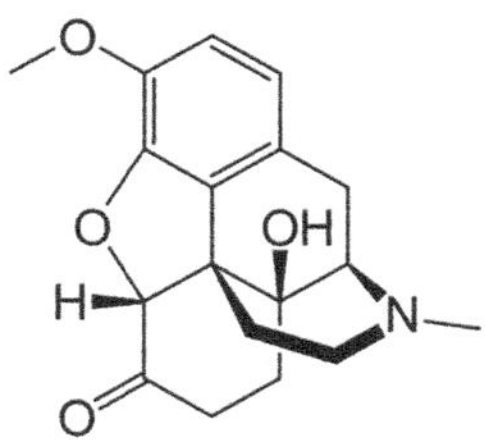

Chemical structure: $C_{17}H_{19}NO_4$

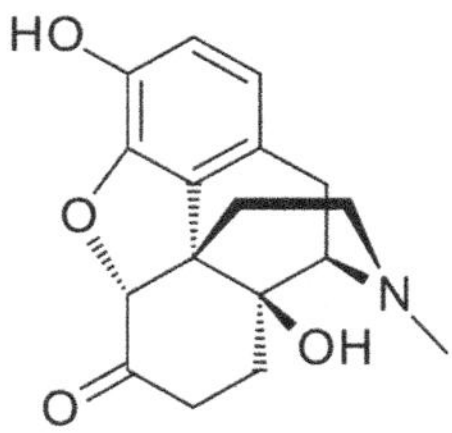

4 Hallucinogens

4.1 *Marijuana*

It is the most common drug used for hallucination (Fig. 7). In addition, it is a common addictive drug worldwide and is highly used in high schools nowadays (Koob et al. 2014; Feldman 2011).

Chemical structure: $C_{21}H_{30}O_2$

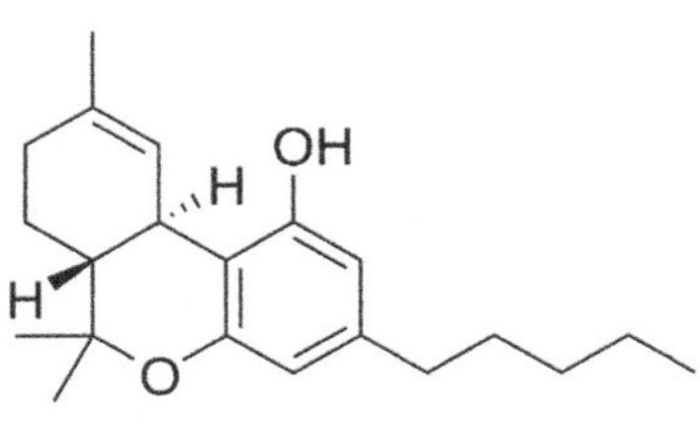

Fig. 7 Positive and negative effects of marijuana

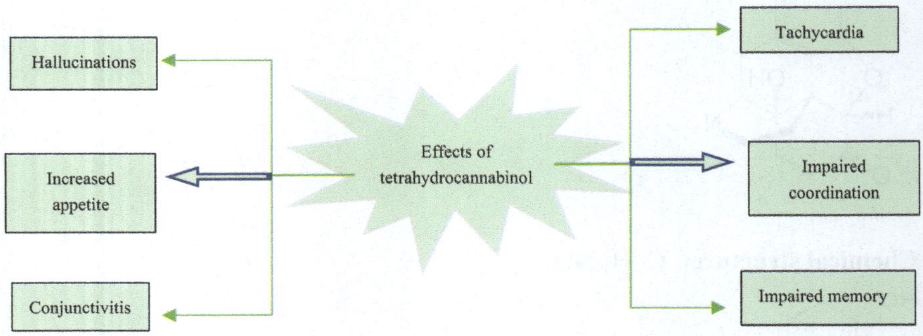

Fig. 8 Effects of tetrahydrocannabinol

The properties of marijuana differ from one to another, but they generally cause euphoria in addition to feeling well (Fig. 8).

There are obvious risks associated with long-term, heavy chronic users of marijuana (Volkow et al. 2014). "Although marijuana does not seem to produce addiction by itself, there is a possibility that marijuana has similarities with other drugs such as cocaine and heroin in how they affect the brain. In addition, there is evidence that heavy use at least temporarily decreases the production of the male sex hormone testosterone, potentially affecting the sexual activity and sperm count" (Volkow et al. 2014).

Medical uses: "Marijuana has several medical uses; it can be used to prevent nausea from chemotherapy, treat AIDS symptoms, relieve muscle spasms for people with spinal cord injuries, and it may be helpful in the treatment of Alzheimer's disease" (Bridgeman and Abazia 2017).

4.2 MDMA and LSD

3,4-Methylenedioxymethamphetamine (MDMA), also known as ecstasy or Molly. It is considered a hallucinogen (National Center for Biotechnology Information 2021).

MDMA enhances feelings of increased energy and euphoria and increased empathy and social life interactions (Kirkpatrick et al. 2014). In a pure form known as Molly, its use is often associated with raves and music festivals, and it has led to a rising number of overdoses and even deaths. There are a lot of data that have been studied.

Chemical structure: $C_{11}H_{15}NO_2$

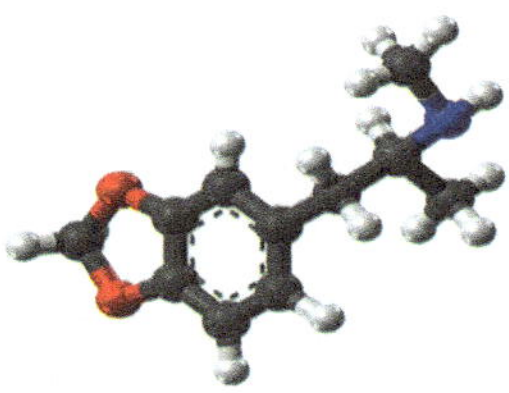

LSD is structurally similar to serotonin that produces significant hallucinations (Fuentes, 2019).

Chemical structure: $C_{20}H_{25}N_{3O}$

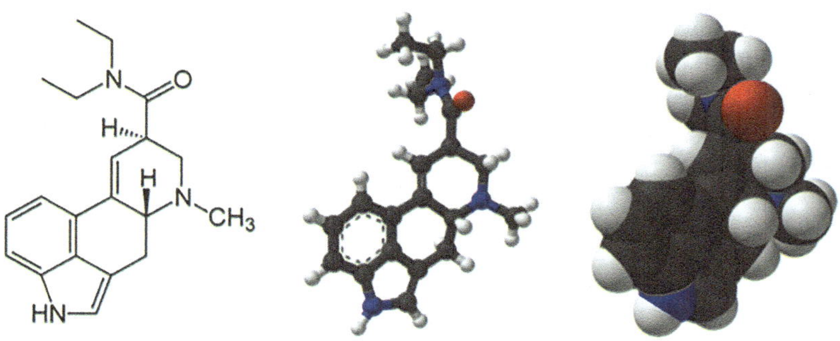

Multiple Choice Questions

1. Which drug is most likely responsible for behavioral excitation, aggressiveness, paranoia, and hallucinations. Also, it can cause mydriasis and elevated HR and BP?
   (a) Marijuana
   (b) Fentanyl
   (c) Alcohol
   (d) Amphetamine

2. Morphine is used clinically in?
   (a) Anxiety
   (b) Depression
   (c) Pulmonary congestion
   (d) Colon cancer
   (e) Alcohol withdrawal symptoms

3. Which one of the following results from methadone ingestion?
   (a) Inhibition of aldehyde dehydrogenase
   (b) Serotonin
   (c) Vomiting
   (d) Formaldehyde
   (e) Dysuria

4. Not a side effect for opioids:
   (a) Diarrhea
   (b) Meiosis
   (c) Sedation
   (d) Respiratory depression

5. Barbiturates work as:
   (a) Dopamine agonist
   (b) CNS stimulants
   (c) GABA agonist
   (d) 5HT antagonist

Answers

1. (d)
2. (c)
3. (d)
4. (a)
5. (c)

References

Ahmed S, Zhou Z, Zhou J, Chen SQ. Pharmacogenomics of drug metabolizing enzymes and transporters: relevance to precision medicine [published correction appears in Genomics Proteomics Bioinformatics. 2018 Apr 21]. Genomics Proteomics Bioinformatics. 2016;14(5):298–313. https://doi.org/10.1016/j.gpb.2016.03.008.

Anthenelli RM, Schuckit MA. Affective and anxiety disorders and alcohol and drug dependence: diagnosis and treatment. J Addict Dis. 1993;12:73–87.

Baldo BA. Opioid analgesic drugs and serotonin toxicity (syndrome): mechanisms, animal models, and links to clinical effects. Arch Toxicol. 2018;92(8):2457–73.

Beaudoin FL, Merchant RC, Clark MA. Prevalence and detection of prescription opioid misuse and prescription opioid use disorder among emergency department patients 50 years of age and

older: performance of the prescription drug use questionnaire, patient version. Am J Geriatr Psychiatry. 2016;24(8):627–36.
Bridgeman MB, Abazia DT. Medicinal cannabis: history, pharmacology, and implications for the acute care setting. P T. 2017;42(3):180–8.
Cami J, Farré M. Drug addiction. N Engl J Med. 2003;349(10):975–86.
Clark MA, Harvey RA, Finkel R, Rey JA, Whalen K. Pharmacology, Fifth Edition. Philadelphia: Lippincott Williams & Wilkins; 2011.
Feldman RS. Understanding psychology. 10th ed. Amherst: University of Massachusetts; 2011.
Fredheim OM, Moksnes K, Borchgrevink PC, Kaasa S, Dale O. Clinical pharmacology of methadone for pain. Acta Anaesthesiol Scand. 2008;52(7):879–89. https://doi.org/10.1111/j.1399-6576.2008.01597.x.
Grant BF. Prevalence and correlates of alcohol use and DSM-IV alcohol dependence in the United States: results of the National Longitudinal Alcohol Epidemiologic Survey. J Stud Alcohol. 1997;58(5):464–73.
Grissinger M. Keeping patients safe from methadone overdoses. Pharm Therap. 2011;36(8):462–6.
Imani F, Motavaf M, Safari S, Alavian SM. The therapeutic use of analgesics in patients with liver cirrhosis: a literature review and evidence-based recommendations. Hepat Mon. 2014;14(10):e23539. https://doi.org/10.5812/hepatmon.23539. Published 2014 Oct 11.
Kirkpatrick M, Lee R, Wardle M, et al. Effects of MDMA and intranasal oxytocin on social and emotional processing. Neuropsychopharmacology. 2014;39:1654–63. https://doi.org/10.1038/npp.2014.12.
Koob GF, Volkow ND. Neurocircuitry of addiction. Neuropsychopharmacology. 2010;35(1):217–38.
Koob GF, Arends MA, Le Moal M. Drugs, addiction, and the brain. New York: Academic Press; 2014.
Lindesmith AR. Addiction & opiates. New York: Routledge; 2017.
National Center for Biotechnology Information. PubChem compound summary for CID 5288826, morphine. 2021. Retrieved October 13, 2021 from https://pubchem.ncbi.nlm.nih.gov/compound/Morphine.
Prus A. Drugs and the neuroscience of behavior: an introduction to psychopharmacology. 2nd ed. Thousand Oaks: SAGE Publications; 2018.
Schiller EY, Goyal A, Mechanic OJ. Opioid overdose. In: StatPearls. Treasure Island: StatPearls Publishing; 2021.
Söderberg Löfdal KC, Andersson ML, Gustafsson LL. Cytochrome P450-mediated changes in oxycodone pharmacokinetics/pharmacodynamics and their clinical implications. Drugs. 2013;73:533–43. https://doi.org/10.1007/s40265-013-0036-0.
Su SS, Larison C, Ghadialy R, et al. Substance use among women in the United States. SAMHSA analytic series A-3. Rockville: Substance Abuse and Mental Health Services Administration; 1997.
Volkow ND, Baler RD, Compton WM, Weiss SR. Adverse health effects of marijuana use. N Engl J Med. 2014;370(23):2219–27. https://doi.org/10.1056/NEJMra1402309.
Wagner FA, Anthony JC. From first drug use to drug dependence: developmental periods of risk for dependence upon marijuana, cocaine, and alcohol. Neuropsychopharmacology. 2002;26(4):479–88.
Wang D. Opioid medications in the management of chronic abdominal pain. Curr Pain Headache Rep. 2017;21(9):40. https://doi.org/10.1007/s11916-017-0640-x. Published 2017 Aug 8.
Weil A. The natural mind: an investigation of drugs and the higher consciousness. Boston: Houghton Mifflin Harcourt; 1998.
Yilmaz B, Konar V, Kutlu S, et al. Influence of chronic morphine exposure on serum LH, FSH, testosterone levels, and body and testicular weights in the developing male rat. Arch Androl. 1999;43(3):189–96. https://doi.org/10.1080/014850199262481.

The Altered States of Consciousness

Mustafa Hayder Kadhim

1 Introduction

Thus, there are those times when our mind seems to split into two kinds of consciousness.

Recently, researchers started studying consciousness varied by hypnosis and drugs. Thus, psychology was regaining the definition of consciousness (Fromm 2017).

Now, there are more research to understand the biology of consciousness. For example, evolutionary psychologists consider that consciousness must offer a seed and soil advantage. Another theory suggests that consciousness enhances our survival by anticipating how we live in the future and reading people's minds.

Even so, the brain leaves us with a lot of complex problems. For example, it creates the awareness of ourselves, the pain of a headache, the feeling of fight and flight (Fromm 2017).

2 Cognitive Neuroscience

It is the science that deals with cognition, and it involves many functions as in Fig. 1.

When researchers asked this young woman to imagine playing tennis or going to university, fMRI scans showed brain activity as healthy volunteers (Fig. 2).

Despite these advancements, much disagreement remains. For example, some researchers suggest that consciousness arises from specific neuronal activity. Others suggest that the whole activity of the neurons produces consciousness.

M. H. Kadhim (✉)
University of Kufa, College of Medicine, Najaf, Iraq

H. T. Hashim, A. Alexiou (eds.), *The Psychology of Consciousness: Theory and Practice*, https://doi.org/10.1007/978-3-030-90692-4_7

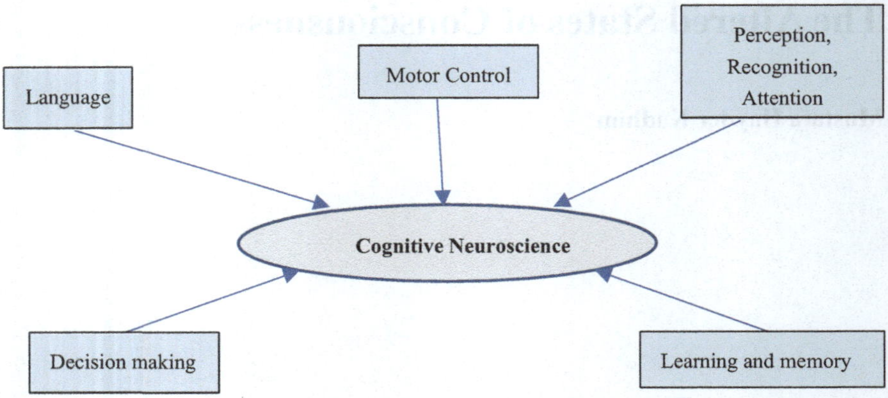

Fig. 1 Cognitive neuroscience, the interdisciplinary trial of the brain excitation integrated with cognition, including memory, language, and thinking

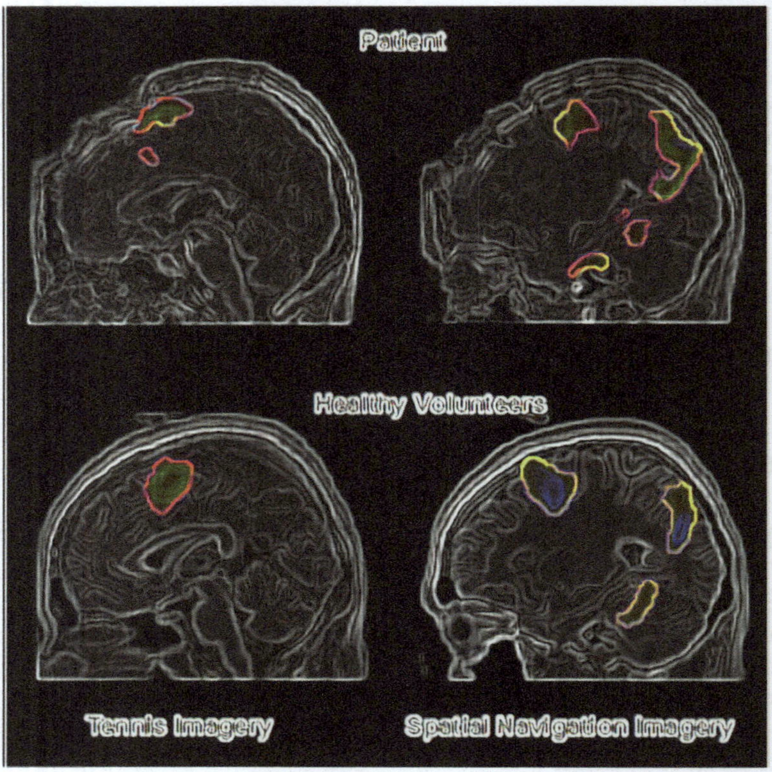

Fig. 2 Evidence of awareness

3 Dual Processing

We are aware of just little about our unconscious mind. Perception, memory, thinking, language, and attitudes all work based on two levels: (Preller et al. 2019).

- Conscious (highly initiative mind)
- Unconscious and automatic (low initiative mind)

That's what we mean when we say "dual processing."

3.1 The Two-Track Mind

A young woman who had brain damage affecting the visual part could not recognize objects and differentiate them visually. But she was only partially blind. So she was asked to handle an object; she could adequately do that (Fig. 3).

Scientists were wondering how this happens. As all of us know, we have only one visual system (Cofré et al. 2020).

Also, the researchers have found that the reverse damage leads to the opposite symptoms. It means a patient may see objects but have difficulty in handling them.

3.1.1 Selective Attention

When you sense ten things by your five senses, you will process only a few of them but not all of them. Yet your unconscious mind will take a process for the other things that the conscious mind has not focused on.

For example, when reading this theory, you are not aware of the watch on your hand or wearing a shoe. When you pay attention to these things, you have been blocking from the awareness coming from the conscious mind (Cofré et al. 2020).

As you have read this example, you may be confused! I know it is genuinely complicated that we have all of these integrative systems in our brains, but we are unaware of them.

Another simple example of selective attention is when four people say your name simultaneously, your mind will focus on one of them and ignore the others (Fig. 4). Also, imagine hearing two conversations simultaneously and being asked to remember both of them; you will not perceive what is said in the other conversation. Considering selective attention and accidents, cell phone use and driving accidents are typical examples. That's why you are not allowed to use a cell phone while driving (Weinel 2018).

Fig. 3 Two-track mind

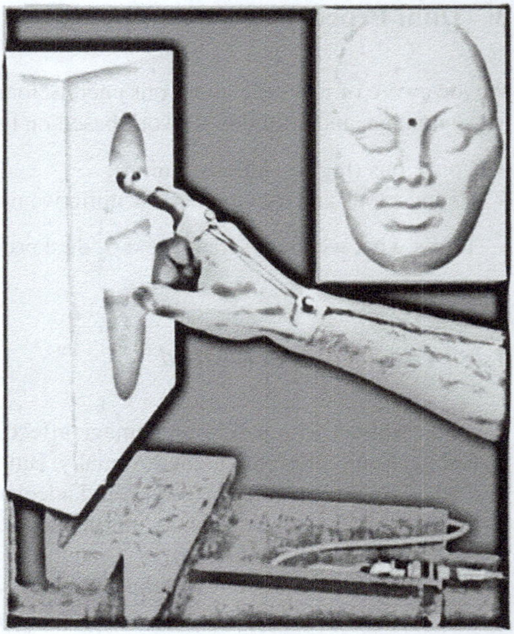

Fig. 4 Selective attention

3.1.2 Selective Inattention and Inattentional Blindness

A somewhat similar idea of what we have talked about in selective attention (Johnson and Kaplan 2019) (Fig. 5).

In other experiments, there is also a blindness to change (Fig. 6). For example, it happened when a man asked a construction worker to direct him to a specific place, and then the construction worker was replaced by another one, but it was unnoticed by the man who asked (Wittmann 2018).

Fig. 5 An example of inattentional blindness

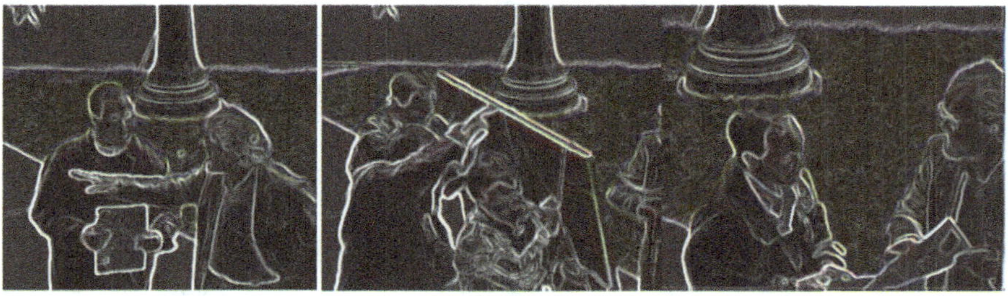

Fig. 6 An example of change blindness

Out of sight, out of mind. But moreover, change deafness can happen (Fig. 7).

A common form of selective inattention happens in media, politics, and various aspects of our lives. An astonishing form called choice blindness (Fig. 8) was invented by a Swedish researcher, "Petter Johansson," and his assistants. In his example, he showed 2 female faces for 5 s to 120 volunteers and asked them which one of those 2 women they found more attractive? Then, the researcher used a sleight of hand (sleight of hand is part of the experimental method) to switch the photo that were handed to the volunteers who did not pick. The volunteers were asked to justify why they had chosen the photo and also justify the choice they haven't made. They were in a weird situation, and they did not noticed that (Holoyda 2020) (Fig. 9).

Selective attention and inattention can even affect your sleep (Fig. 10) and what you have experienced in your dreams.

The other aspects of altered states of consciousness (Fig. 11) can be summarized in the following topics:

(a) Sleep and dreams
(b) Hypnosis
(c) Drugs and consciousness
(d) Meditation and yoga

Fig. 7 An example of change deafness

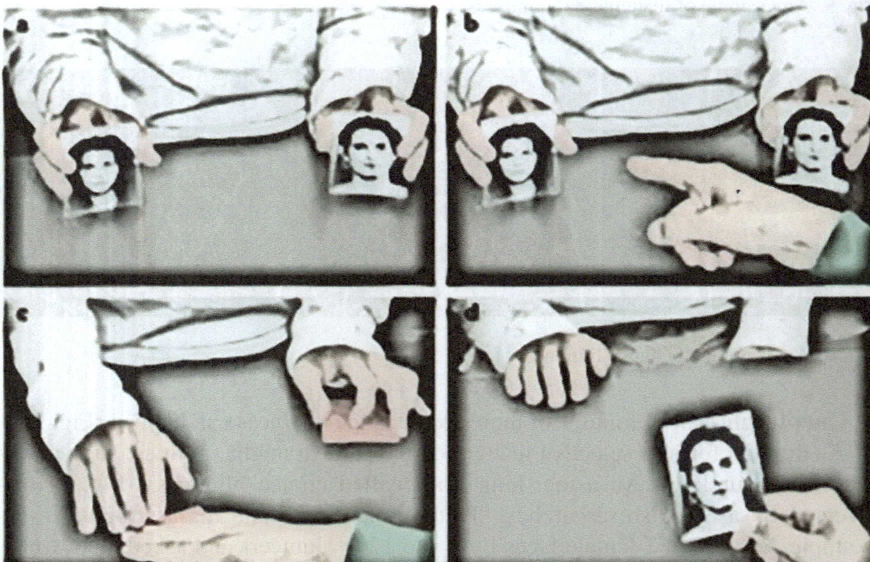

Fig. 8 An example of choice blindness

Sleep and Dreams

Sleep is a basic, natural, and easily reversible periodic phenomenon characterized by a diminished level of awareness, reduced response to the sensory input, and executive motor output with minimal interaction of the body with the external world. More details on this topic have already been discussed in chapter "Sleep and Dreaming
".

Fig. 9 Pop-out phenomenon

Fig. 10 Selective attention and inattention may affect your sleep

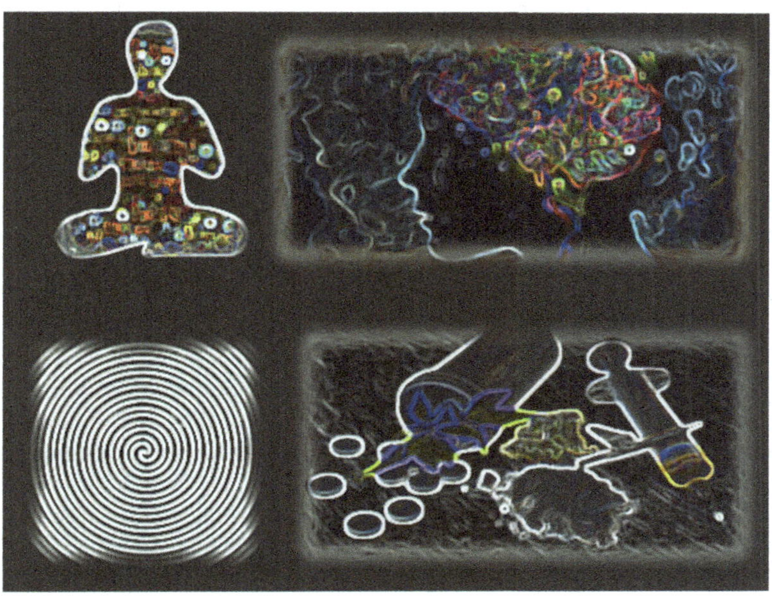

Fig. 11 The other aspects of altered states of consciousness

Hypnosis

Hypnosis is a divided state of consciousness. It can be used in managing pain and treating many disorders or health issues. More details on this topic will be discussed in chapter "Hypnosis and Consciousness".

Drugs and Consciousness

There are many drugs that have an effect human's consciousness starting from altering its states, addiction to changing the mood. More details on this topic have already been discussed in chapter "Drugs and Consciousness".

Meditation and Yoga

It is defined as a learned technique for refocusing attention and rebooting it, leading to an altered consciousness (Moreira-Almeida and Lotufo-Neto 2017).

Meditation affects the mind through different techniques, but it has the same goal of relaxing and refocusing attention. You can do it while exercising, sitting, and lying down (Fig. 12). It would be best if you think of nothing for a while (Schmidt and Berkemeyer 2018).

There are various ways to do meditation. Most of it is done sitting in special postures, such as:

- the full or half lotus
- using relaxed cushions or sitting on a couch and the feet bent underneath.

All these positions have the same goal; to reach a spiritual insight that is to relax and to be alert. There is nothing distinct about these positions. In meditation, people need to achieve a certain level of stabilization and avoid two things: either becoming sleepy or being nervous by distracting thoughts or perceptions. The unique postures provide a good base, relaxed state, a straight spine, and deep breathing. So, avoiding these two irregular things is what we have mentioned.

People say that they are feeling fully relaxed after meditation. They claim that it has relieved their stress life event and made them more patient in facing the problems. Barnes, Kleinman, and Travis found that oxygen usage decreases, blood pressure and heart rate decrease, and brain activities change during meditation (Chieffi et al. 2018; Kandeepan et al. 2020).

With more and more training, it is possible to calm our minds and leave the distractive thoughts off. Furthermore, our brains may adapt to cope with life worries and problems and help us to reach the inside happiness (Fig. 13). Although some experiments are surprising, they found that meditation is no more relaxing than the other ordinary relaxed states. Some argue that it even may be more problematic if specific unwanted thoughts keep coming up and you cannot control your mind and emotions. Therefore, they say it is better to do exercise than to meditate.

Fig. 12 Ways of meditation

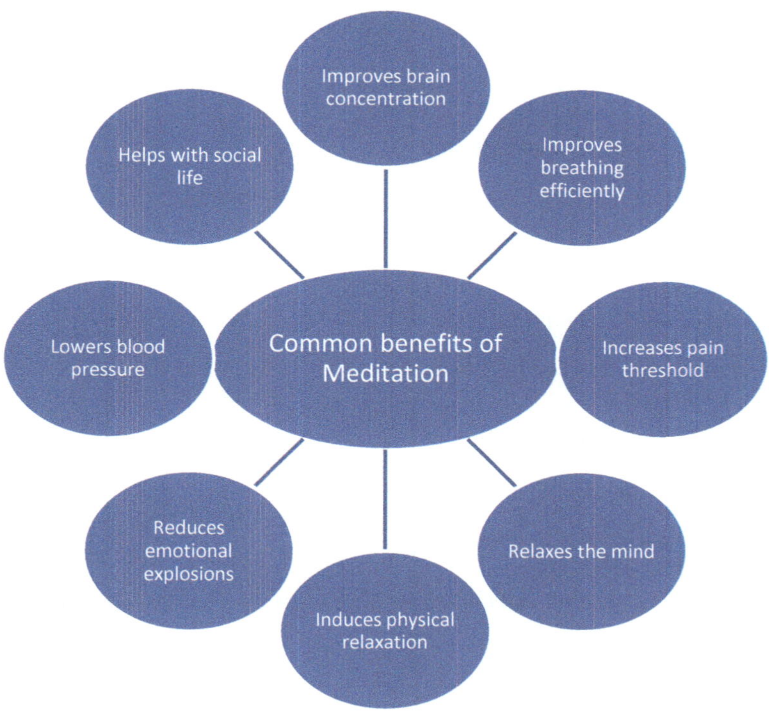

Fig. 13 Benefits of meditation

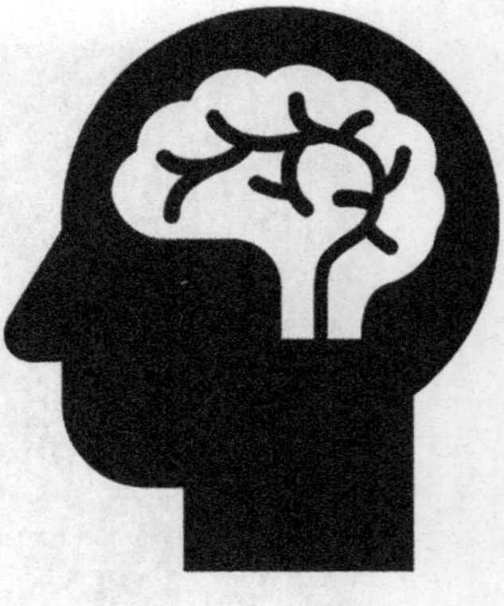

Fig. 14 Meditation may lead to greater spiritual insight

Meditation is practiced in many cultures and religions. It takes various aims and different goals among cultures. Interestingly, we found that people with different cultures and religions seek ways to alter their consciousness to achieve spiritual experience (Fig. 14).

Multiple Choice Questions

1. Which of the following is not a characteristic of meditation?
   (a) Increases pain threshold
   (b) Relaxes the mind
   (c) Elevates the BP
   (d) Improves brain concentration

2. Hypnotherapy has been used for which of the following?
   (a) Cancer
   (b) Cardiac disease
   (c) Respiratory disease
   (d) Musculoskeletal problems

3. Hypnosis has all of the following, except:
   (a) Two-track mind
   (b) Biological influences
   (c) Psychological influences
   (d) Social-cultural influences

4. Selective attention is defined as:
   (a) Psychological disease
   (b) Focusing of mind and conscious awareness on a particular stimulus
   (c) Physiological mind
   (d) None of the above

5. The two-track mind is described as:
   (a) Cultural mind
   (b) Biological influences on the mind
   (c) When you can see objects and handle them at the same time
   (d) When the patient can't see objects but can handle them
6. Selective attention and inattention can affect which of the following:
   (a) Stress
   (b) Diseases
   (c) Eye function
   (d) Sleep
7. Cognitive neuroscientists can activate specific patterns and see your brain thinking, depending on which of the following:
   (a) Basal ganglia
   (b) Thalamus
   (c) Brain stem
   (d) Cortex
8. All of the following are classified as spontaneous states of consciousness, except:
   (a) Dreaming
   (b) Drowsiness
   (c) Meditation
   (d) Dreaming
9. Choice blindness is:
   (a) Selective attention
   (b) Pop-out phenomenon
   (c) Hypnosis
   (d) Meditation
10. Psychological influences of hypnosis include all of the following, except:
    (a) Focused attention
    (b) Expectation
    (c) Dissociation between normal sensation and conscious
    (d) Distinctive brain activity

Answers

1. (c)
   Meditation helps reduce blood pressure and increases pain threshold, relaxes the mind, and improves brain concentration.
2. (c)
   It has been used as a treatment for asthma. Also, it is used for headaches and stress-related skin disorders.
3. (a)
   Two-track mind is a theory that describes the mind's two levels, and it is unrelated to hypnosis.
4. (b)
   Selective attention is the focus of mind and conscious awareness on a particular stimulus.
5. (d)
   It describes as the mind's two levels.
6. (d)
   Selective attention and inattention can even affect your sleep and what you have experienced in your dreams.
7. (d)
   Depending on your cortical function and consciousness, they can activate specific cortical patterns and see your brain thinking in limited ways.
8. (c)
   Meditation is classified as a psychologically induced state of consciousness.
9. (d)
   Choice blindness is an example of selective inattention which is used in politics, media, and various aspects of our lives.
10. (d)
    Distinctive brain activity is one of the biological influences of hypnosis.

References

Chieffi IV, et al. Mind, brain and altered states of consciousness. Acta Medica. 2018;34:357.

Cofré R, et al. Whole-brain models to explore altered states of consciousness from the bottom up. Brain Sci. 2020;10(9):626.

Fromm E. The nature of hypnosis and other altered states of consciousness: an ego psychological theory 1. In: Hypnosis. New York: Routledge; 2017. p. 81–104.

Holoyda B. Psychedelic psychiatry: preparing for novel treatments involving altered states of consciousness. Psychiatr Serv. 2020;71(12):1297–9.

Johnson EL, Kaplan PW. Clinical neurophysiology of altered states of consciousness: encephalopathy and coma. Handbk Clin Neurol. 2019;161:73–88.

Kandeepan S, et al. Modeling an auditory stimulated brain under altered states of consciousness using the generalized Ising model. Neuroimage. 2020;223:117367.

Moreira-Almeida A, Lotufo-Neto F. Methodological guidelines to investigate altered states of consciousness and anomalous experiences. Int Rev Psychiatry. 2017;29(3):283–92.

Preller KH, et al. Effective connectivity changes in LSD-induced altered states of consciousness in humans. Proc Natl Acad Sci. 2019;116(7):2743–8.
Schmidt TT, Berkemeyer H. The altered states database: psychometric data of altered states of consciousness. Front Psychol. 2018;9:1028.
Weinel J. Inner sound: altered states of consciousness in electronic music and audio-visual media. Oxford: Oxford University Press; 2018.
Wittmann M. Altered states of consciousness: experiences out of time and self. Cambridge: MIT Press; 2018.

Hypnosis and Consciousness

Hashim Talib Hashim and Mustafa Ahmed Ramadhan

1 Hypnosis

It is defined as a state of divided consciousness (Fig. 1). The hypnosis, when it is applied, will split the consciousness into two levels as follows:

- One level is to obey the commands that the hypnotist says.
- The other level is to be aware of the environment around them, and they may be aware that someone is talking to them.

Imagine that you are about to be hypnotized; you are sitting on a chair and looking at a fixed spot on the wall. The hypnotist asks you to relax and suppose that you are tired, your eyes become heavy more and more, you are becoming more deeply relaxed, and you feel like you are dead (Orne 2017).

When hypnotic induction occurs, you will experience hypnosis. You cannot open your eyes even if you try to open them; you cannot control yourself; you will obey anything you have been told. For example, if you are told to forget the number 4, you will count 6 fingers in your hands.

Can we consider hypnosis as really an altered state of consciousness? We will discuss hypnosis through its different aspects to answer this question.

Hypnosis is a technique in which one person responds to suggestions provided by another person (Hypnotist) for creative encounters involving changes in vision, memory, and voluntary regulation of behavior by one person (Subject).

Hypnotized people may be ignorant of pain; they hear voices that are not there. They are unable to recall the things that happened to them when they were hypnotized. They cannot see details distinctly in their field of vision and make comments after hypnosis has been stopped, without being conscious of what they are doing or why (Bowers and Bowers 2017)

H. T. Hashim (✉) · M. A. Ramadhan
University of Baghdad, College of Medicine, Baghdad, Iraq

H. T. Hashim, A. Alexiou (eds.), *The Psychology of Consciousness: Theory and Practice*, https://doi.org/10.1007/978-3-030-90692-4_8

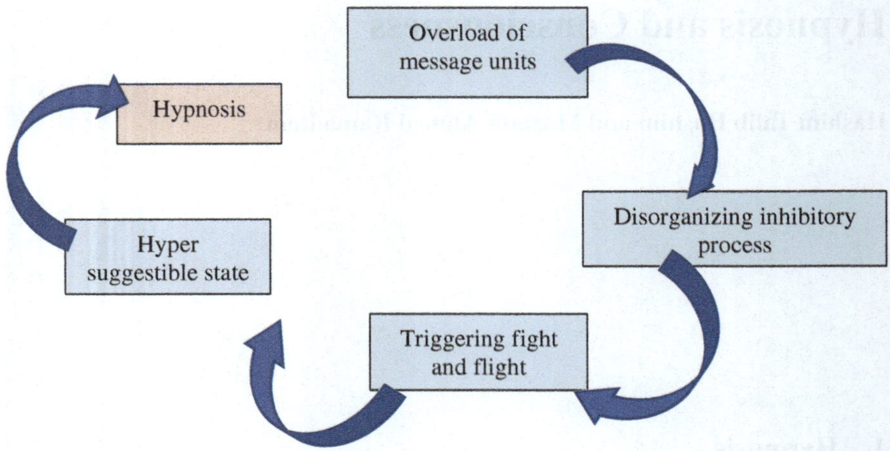

Fig. 1 Induction process of hypnosis

Both encounters are associated with a degree of subjective belief that borders on delusion and an experience of involuntariness that borders on compulsion in the classic example.

Subjects respond to recommendations for creative interactions in hypnosis that may include cognitive perception, memory, and behavior changes. However, among those incredibly hypnotizable topics, these phenomena arise more intensely (Valentine et al. 2019).

While most individuals may undergo hypnosis to at least some degree, the most extreme hypnosis phenomena are typically found in those that count as representing changes in consciousness. Thus, those "hypnotic virtuosos" make up the top 10–15% of the hypnotizable distribution.

A great deal of hypnosis study thus requires a priori collection of subjects who are highly hypnotizable, to the exclusion of those with low and intermediate hypnotizable. An alternative is a hybrid system in which subjects stratified for hypnotizable are all subjected to the same experimental manipulations, and hypnotizable subjects' reactions are compared to others subjected to the same experimental manipulations. Hypnosis is insusceptible. In either case, hypnotizable estimation is important for research into hypnosis; there is no point in analyzing hypnosis in people who can't feel it.

Any psychiatric clinicians assume that if only the hypnotist follows the correct path, nearly all can be hypnotized, but no studies favor this point of view. Similarly, some experts suggest that hypnotizable can be increased by cultivating constructive hypnosis beliefs, motives, and aspirations although there is also evidence that these measures are closely related to enforcement.

The multifaceted essence of hypnosis itself is part of the issue. Hypnosis requires improvements in conscious thought, memory, and actions, to be sure, but these improvements often arise following explicit recommendations given to the subject by the hypnotist.

Thus, hypnosis is simultaneously a state of radical neurological improvement, including essential vision, memory, thinking processes, and social interaction in which hypnotic and subject come into play. Under a broader sociocultural framework, together with a particular reason, a fully adequate, systematic hypnosis theory would attempt comprehension in cognitive and behavioral terms.

The inter-individual rivalry that is part of alterations and parcel of science as a collective enterprise frequently causes scholars to write those mutually exclusive mechanisms were alterations of consciousness and social power-which they clearly are not (Shor 2017).

In the case of posthypnotic amnesia and hypnotic analgesia, hypnotic anesthesia, and negative hallucinations, it appears that hypnotized participants cannot become conscious of thoughts and memories that would usually be expressed in phenomenal consciousness.

On the other hand, hypnosis has often been proposed to have the opposite ability to allow participants to become aware of thoughts and experiences that are not ordinarily available through active introspection.

In hypnotic hypermnesia, for example, participants receive suggestions that they may be able to recall events they have forgotten (Terhune et al. 2017).

Automatic processes are carried out automatically in a reflex-like manner as a first approximation, whereas regulated processes are carried out intentionally.

The posthypnotic suggestion is probably the most dramatic example of the apparent automaticity of hypnosis. The participants react to a suggestion offered when they were already hypnotized after the end of hypnosis.

Any proposed sensation that might arise during hypnosis may also arise posthypnotically if the subject is hypnotizable enough. For this explanation, for many conceptions of hypnosis as an altered state of mind, posthypnotic persuasion has often been troublesome because the phenomenon happens after the hypnotic state has been ostensibly ended. As far as we can see, although they respond to the posthypnotic recommendation, participants do not re-enter hypnosis. Centered on the unification of four factors, hypnosis may be characterized as an altered state of consciousness: induction technique, subjective perception, accessible behavior, and psychophysiological indices—including neural correlates of hypnotic persuasion discovered by brain imaging.

Hypnotic induction requires focused concentration and creative engagement to the point that it feels true to what is being thought. The clinician and patient create a hypnotic reality through the use and approval of advice.

For a medical reason, hypnosis may be used as a meditative condition, where one may learn to enter intentionally and actively. Then, suggestions for the desired effect are offered either orally or using imagery. For example, this may be to reduce discomfort by getting access to calmness and relaxation, helping to control drug side effects, or helping to relieve pain or other symptoms.

Hypnosis is typically a calming experience, based on the advice offered, and may be very helpful for a tense or anxious patient (Williamson 2019).

Fig. 2 Hypnosis can assist people in accepting and experiencing what can be done by them

Hypnosis does not make the impossible possible (Williamson 2019). Instead, it can assist people in accepting and experiencing what can be done by them. Since civilization has existed, hypnotic states have been used for treatment (Fig. 2).

It was possible to see the right brain as the most emotional, imaginative part of us that deals with metaphors and pictures and may be seen as our unconscious mind. However, it is still impossible to state that we are not angry or nervous because words are not the right brain's language.

They work at an emotional rather than logical level while patients are highly nervous, and one should activate and direct their imaginative creativity to what is beneficial for them. To construct potential catastrophic situations, nervous people use their imagination, which creates still more fear and thus more dopamine, which can spiral into hysteria (Jensen et al. 2017).

Patients can find that their feelings are overloaded. If health providers can stimulate their energy, guide their creativity to feel relaxed, or re-encounter a meaningful experience or behavior and have positive feedback, they will feel calmer and can cope more (Williamson 2019; Lynn et al. 2020).

There are some similarities shared between hypnosis and meditation. Both are states of concentrated attention, and both can be used to create a profound feeling of relief (Keshmiri et al. 2020).

1.1 *Hypnosis Experience*

As we said, hypnosis is a social interaction where the hypnotist asks the subject to experience certain thoughts, behaviors, and feelings.

So, we can say that we are all open to suggestions. For example, when the hypnotist asks the hypnotized person to stand and close his or her eyes, and tells he or she is swinging now; surprisingly, the subject will sway slightly. Some researchers say that people with or without hypnosis will respond differently.

The hypnotist supposes various experiences ranging from easy to difficult after a hypnotic induction (ranging from moving your legs to opening your mouth, opening your eyes and imagining what the hypnotist said). For example, Barnier and McConkey had said that the people who did not react to the smell of a bottle of ammonia are those who deeply went into the hypnosis and experienced the imaginary sight.

So, there is a difference between people in hypnosis experience depending on your hypnotic susceptibility.

In conclusion, anyone can experience hypnosis by themselves when integrating the inward and outward imaginary ability. If you expect to be hypnotized, you will experience it. Imagine that you are sitting at the table after hypnotic induction, and the hypnotist asks you to draw a circle; you will write it without the ability to eject it.

1.2 Hypnosis Enhancement and Its Recalling of Forgotten Events

For instance, can hypnosis help elderly people to remember their primary school classmates, to discover their fears and details of a crime? Can it be used in criminal investigations (Franz et al. 2020)?

Also, it is an inaccurate statement to say that hypnosis can recall accurate memories of people since birth. Johnson and Hauck proved this in 1999. Furthermore, there is no supposed ability to relive your own childhood experiences. They feel like real children and speak like them, but there is not any change in adult brain waves in MRI. Now, it is banned that you are making evidence from witnesses who have been hypnotized because the hypnotist can create pseudo-memory to the hypnotized subject. For example, by asking "*did you feel that you are entering someone's home .... stealing their properties .... killing someone,*" so you can plant false information and create pseudo-memory.

Sometimes, hypnosis may be used in the wrong way that the hypnotist induces the subject to do a scary deed. For example, asking the subject to stand up, take a knife or any other thing, and throw it onto a researcher. Surprisingly, some people do not remember their acts and deny doing such things.

1.3 Hypnosis as Divided Consciousness

There is a conflict and controversial debate about this theory. However, most hypnosis researchers suggest that regular social and cognitive neuroscience play an essential part in hypnosis.

The researcher, Perugini, in 1998, demonstrated that: "For one thing, hypnotized subjects will sometimes carry out suggested behaviors on cue, even when they believe no one is watching."

In an experiment, people whether hypnotizable or not were asked to say the color of letters (the word GREEN in red color to create a little bit of conflict). Unhypnotized people were slowed from this conflicting idea. At the same time, hypnotized people were quickly slowed by this conflict and were able to solve this conflict (Brain areas involved in decoding words and revealing the conflict remained inactive). Famous researcher Ernest Hilgard (1986–1992) believes in social influence theory and divided-consciousness theory. It means that hypnosis involves social influence theory and a defined theory of dissociation—a split between different levels of consciousness. Hilgard believed that splitting consciousness is a part of everyday life events. For example, sometimes, we paint or write a poem while listening to a lecture.

Hypnosis also may lead to pain relief. A PET scan supported this. Thus, hypnosis blocks our attention to painful stimuli, but it cannot block sensory input in the brain cortex (Shalbaf et al. 2020).

1.4 Uses of Hypnosis

Many clinical and psychological applications exist for this technique in treatment, research, and diagnosis (Fig. 3). We will mention them in the following points:

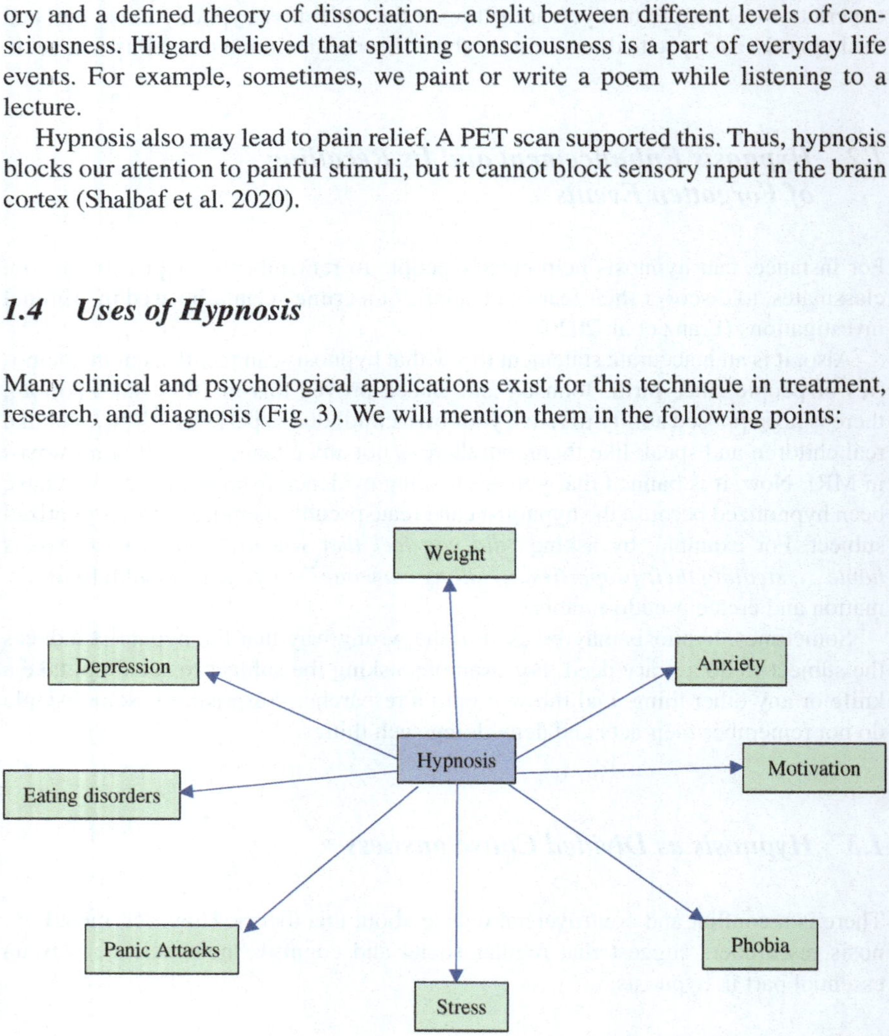

Fig. 3 Hypnosis uses

1.4.1 Hypnotherapy

The application of hypnosis in psychotherapy is hypnotherapy. It is useful for doctors, psychiatrists, and those who are licensed (Levitt and Chapman 2017).

It has been used in a range of forms, with varying results, such as:

- Addiction
- Psychotherapy
- Relaxation
- Sports performance
- Weight loss
- Soothing anxious surgical patients
- Phobias

1.4.2 Irritable Bowel Syndrome

For the prevention of irritable bowel syndrome, hypnotherapy has been examined. In the National Institute for Wellbeing and Clinical Excellence guidelines released by UK health services, hypnosis for IBS has gained modest support. It has been applied as an aid or substitutes for chemical anesthesia and has been researched as a means to soothe skin ailments. However, the topic is still under development and needs more studies and research to be approved (Gruenewald et al. 2017).

1.4.3 Pain Management and Cancer Supporter

In minimizing the anxiety of cancer therapy, hypnosis effectively reduces sufferings due to cancer and other chronic conditions. With hypnosis, fatigue and other symptoms linked to incurable conditions can also be treated (Gruenewald et al. 2017).

Hypnosis has also been used during oral surgery and associated pain control regimens as a pain-relieving strategy (Gruenewald et al. 2017).

Multiple Choice Questions

1. Hypnosis is defined as:
   (a) A state of divided consciousness
   (b) A state of sleepiness
   (c) A state of being unconscious
   (d) A state of dreaming
2. ……………… are carried out automatically in a reflex-like manner as a first approximation?
   (a) Nervous reflex
   (b) Psychiatric interview
   (c) Autonomic associations
   (d) Conscious–unconscious relationships
3. Hypnosis may be used as a meditative condition, where one may learn to enter intentionally and actively.
   (a) For criminal reasons
   (b) For resting reasons
   (c) For acting reasons
   (d) For a medical reason
4. For cancer patients, what does hypnotherapy do?
   (a) It relieves pain
   (b) It is a palliative therapy to reduce pain
   (c) It reduces metastasis
   (d) It limits the treatment options
5. Hypnosis has also been used during oral surgery and associated pain control regimens as a (an):
   (a) Pain-relieving strategy
   (b) Anesthetic strategy
   (c) Curative strategy
   (d) Calming strategy

Answers

1. (a)
2. (c)
3. (d)
4. (b)
5. (a)

References

Bowers KS, Bowers PG. Hypnosis and creativity: a theoretical and empirical rapprochement. In: Hypnosis: research developments and perspectives. New York: Routledge; 2017. p. 255–92.

Franz M, et al. Suggested deafness during hypnosis and simulation of hypnosis compared to a distraction and control condition: a study on subjective experience and cortical brain responses. PLoS One. 2020;15(10):e0240832.

Gruenewald D, Fromm E, Oberlander MI. Hypnosis and adaptive regression: an ego-psychological inquiry. In: Hypnosis: research developments and perspectives. New York: Routledge; 2017. p. 495–510.

Jensen MP, Jamieson GA, Lutz A. New directions in hypnosis research: strategies for advancing the cognitive and clinical neuroscience of hypnosis. Neurosci Conscious. 2017;3:1–14.

Keshmiri S, Alimardani M, Shiomi M, Sumioka H, Ishiguro H, Hiraki K. Higher hypnotic suggestibility is associated with the lower EEG signal variability in theta, alpha, and beta frequency bands. PLoS One. 2020;15(4):e0230853. https://doi.org/10.1371/journal.pone.0230853.

Levitt EE, Chapman RH. Hypnosis as a research method. In: Hypnosis: research developments and perspectives. New York: Routledge; 2017. p. 85–114.

Lynn SJ, et al. Hypnosis and health psychology. In: The Wiley encyclopedia of health psychology. New York: Wiley; 2020. p. 257–63.

Orne MT. On the simulating subject as a quasi-control group in hypnosis research: what, why, and how. In: Hypnosis: research developments and perspectives. New York: Routledge; 2017. p. 399–444.

Shalbaf A, et al. Monitoring the level of hypnosis using a hierarchical SVM system. J Clin Monit Comput. 2020;34(2):331–8.

Shor RE. The fundamental problem in hypnosis research as viewed from historic perspectives. In: Hypnosis: research developments and perspectives. New York: Routledge; 2017. p. 15–40.

Terhune DB, et al. Hypnosis and top-down regulation of consciousness. Neurosci Biobehav Rev. 2017;81:59–74.

Valentine KE, et al. The efficacy of hypnosis as a treatment for anxiety: a meta-analysis. Int J Clin Exp Hypn. 2019;67(3):336–63.

Williamson A. What is hypnosis and how might it work? Palliat Care. 2019;12:1178224219826581. https://doi.org/10.1177/1178224219826581.

Artificial Consciousness

Hashim Talib Hashim and Mustafa Ahmed Ramadhan

1 Introduction

In 1992, the word "artificial intelligence" was used scientifically for the first time. Since then, scientists have not agreed on when such an accomplishment is feasible. Every recognized physical structure does not possess the observed properties of the human mind, including unity, representation, and being in relation to each other (Hildt 2019).

The Turing test is the most well-known tool for measuring computer intelligence (Shieber 2004). Therefore, it was also proposed that the suggestion of Alan Turing to mimic not the consciousness of a human adult, but the consciousness of a human child should be taken seriously.

2 Digital Consciousness

It means using the computers to deliver and create a machine consciousness among robots to deliver the same or at least similar functions of the human brain by processing data and codes (Argonov 2014; Searle et al. 1980; Oppy and Dowe 2011; Alexiou et al. 2020) (Fig. 1).

Mathematics and logic may, to a certain degree, describe and evaluate knowledge understood as rationality, and computer science and rtificial intelligence as a whole are primarily focused on mathematics and logic (Fig. 2). However, when it comes to consciousness, mathematics terminates, and theory continues. From the hippocampus down to the genetic code, scholars have sought to locate the origins of consciousness.

H. T. Hashim (✉) · M. A. Ramadhan
University of Baghdad, College of Medicine, Baghdad, Iraq

H. T. Hashim, A. Alexiou (eds.), *The Psychology of Consciousness: Theory and Practice*, https://doi.org/10.1007/978-3-030-90692-4_9

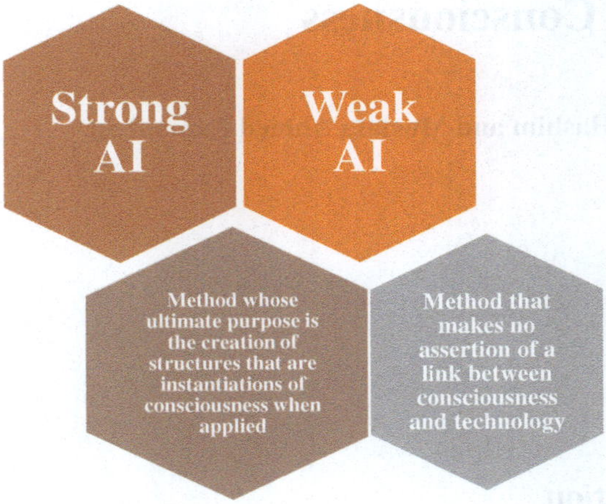

Fig. 1 Strong AI and weak AI

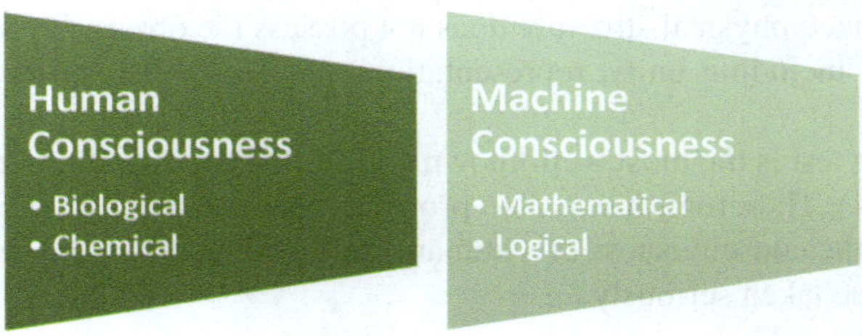

Fig. 2 Human vs. machine consciousness

The structure of consciousness doesn't appear to be solely mechanical or chemical. If it does not only matter, though but something beyond matter, which it appears to be, there is no possibility of artificial consciousness.

Perhaps consciousness will remain transcendental to human awareness, no matter how hard we try. Consciousness itself is the very source of logic. Thus, by applying intuition, it is challenging to grasp consciousness. How does someone comprehend the source through a result of the source? To some point, perhaps, one can understand, but not entirely.

For instance, based on studies on aggressive actions against robots, Kate Darling argues that treating robots more like pets than just objects aligns with our social values (Manzotti and Chella 2018).

While the precise arguments for granting robot privileges vary, similar to these positions, they reflect on the social functions that people give to robots, the interactions, and emotional connections that people develop with robots, or they can communicate with them (Ernst et al. 2021).

2.1 Philosophical and Ethical Aspects of Artificial Consciousness

Several concerns about artificial consciousness and creating similar mind processing capabilities to humans raise many issues to be solved and controlled (Koch 2019). Channeling the philosopher Thomas Nagel, we might conclude that if there is anything like that in a machine, a system is alive. Can the machine have feelings and emotions when dealing with humans? There is no evidence that our intellect and perceptions, rather than any divine ones, are inevitable products of our brain's inherent causal forces.

Therefore, a host of subsidiary mechanisms such as expression, preparation, compensation circuits, and short-term memory buffer storage have access to the signal. What makes us aware is the process of broadcasting this information internationally.

Aware states emerge from the workspace algorithm processing the appropriate sensory inputs, motor outputs, and internal variables linked to memory, inspiration, and anticipation. What awareness is about is global production. Consciousness is just a smart hack away. You can go back to chapters "Brain and Mind" and "Levels of Consciousness" to understand these ideas extensively (Koch 2019).

Although in ascriptions of consciousness there is broad disagreement, there is one thing similar to all of the above; superintelligence is not needed. To show this, the human case itself is necessary. Many hypotheses go even farther, ascribing cognition to beings of rudimentary knowledge or indeed no intelligence in certain instances. Extrapolation of human intelligence in minimally intelligent artificial systems might make it possible to replicate consciousness. Our perception of consciousness must keep pace with the development in artificial intelligence and be extended to it.

As with artificial intelligence, in following artificial consciousness, a distinction may be made between two linked but separate priorities.

Artificial consciousness is mainly concerned with understanding the processes underlying consciousness. Therefore, the tools offered by engineering artificial consciousness, however excellent, are considered to be of theoretical significance only to the degree that they approximate or otherwise illuminate the processes underlying consciousness.

Inside artificial consciousness, more distinctions can be made about the relationship between the hardware used for an artificial consciousness system and the consciousness that it is assumed to hold. Adapting terms from weak artificial consciousness is a method that does not assert a link between consciousness and technology. This will be a use of consciousness awareness technologies similar to computational hurricane simulations in meteorology: such simulations can promote understanding, but no one believes that this is because hurricanes are themselves computational in any substantive sense.

On the other end, strong artificial consciousness is any method whose ultimate purpose is to create structures that are instantiations of consciousness when applied.

A symbolic artificial intelligence approach to powerful artificial consciousness, for instance, ensures that a correctly configured machine is genuinely conscious, aware, experienced, etc. An overlooked field of probability, called lagom artificial consciousness, is between these two extremes. "Lagom" is a Swedish term with no exact English counterpart, meaning "perfection by moderation." Unlike weak artificial consciousness, the lagom artificial consciousness view argues that the simulation relationship holds due to technology and conscious mental phenomena sharing broader explanatory properties.

Unlike strong artificial consciousness, however, lagom artificial consciousness does not assert that instantiating these common properties alone is necessary to instantiate awareness—something else might be needed. Thus, although strong artificial consciousness attempts to discover adequate conditions for consciousness, it would be natural to conclude that lagom artificial consciousness only tries to discover (some of them) necessary conditions for it. But that isn't exactly right. That would be so if only one way of thinking existed.

The argument that the mind is computational comes with different strengths, but the upshot can be used to keep it concise. In the related context, the argument that some computational ideas are part of the interpretation of specific behavioral processes, though much weaker (and thus more difficult to defeat) than the arguments that artificial intelligence opponents typically argue against, is nevertheless intriguing. It means that computer science has a role to play in understanding the mind, and not only in the weak manner in which hurricanes are understood. The equivalent argument for an artificial intelligence approach to artificial consciousness is that some programming principles are part of interpreting some conscious phenomena. A significant difference within artificial consciousness is whether one is directly trying to reproduce/explain human (or biological) consciousness or, more broadly, trying to reproduce/explain consciousness. Either more ambitious or more modest may be the search for generality. Artificial consciousness, for example, generality, would be more ambitious in constitutive science if one sought to explain why the agent is conscious of every possible conscious agent (Vervoort et al. 2021).

And if one's overall aim is to clarify human consciousness, it may help achieve the mission by explaining how a unique artificial machine can become conscious. As Fodor says, "No one has the slightest idea of how aware something material might be. No one knows what it would be like to have the smallest idea of how aware any substance might be."

A known, non-human artificial consciousness would give us something more than a brief understanding of how consciousness might be anything material.

For discriminative accounts of consciousness, comparable arguments may be made. Even if one could not justify why that is so, a discriminatory account of the separate experiential states would benefit an artificial system assumed to be alive, even if it did not have a discriminatory account that applied to the human situation.

This functionalist approach suggests that, in terms of their abstract causal structure, one should provide an overview of mental states in reality. This presumption, at least for those cognitivistically inclined, is not troublesome in the case of thought. It seems logical to conclude that the steps are available to the thinker in an episode of thought; thus, it is usually possible for a thinker to write down certain steps in a finite list. At least some people assume that it is not inconceivable that, with regard to its experiential properties, there may be something that is biologically (and hence behaviorally) similar to you and still separate from you, even to the extent of not getting any memories at all.

These two intuitions are in sharp conflict: the naturalistic intuition is that you fix everything else if you fix the actual problem. On the other hand, the zombic hunch is that you somehow have not resolved the feeling even if you fix it physically (Fig. 3). Thus, an unsatisfactory cognitive dissonance is created by the existence of any of these intuitions.

There are many explanations why one would assume that artificial consciousness cannot be done in the context of engineering or that it cannot lead to an experience of consciousness.

Although the notion that phenomenal cognition cannot be realized in computers seems to be an apparent inference, there are grounds for a more thorough analysis of this issue. Advances in artificial intelligence are growing with immense speed. There will be greater connectivity to more efficient computers that can do more complex computation as software and hardware developments continue to advance (Hromiak 2020).

There are computer programs to perform identification, target tracking, and face recognition.

From simple information gathering to extremely dexterous gestures, logic, and the development of aesthetic experiences, the human mind can perform many amazing acts and calculations (Haladjian and Montemayor 2016).

Attention is significant because it is how the brain selectively stores information derived from sensory and memory inputs (Haladjian and Montemayor 2016).

The focus of the universe will also function on objects that show object-like features, such as balance, symmetry, and collective destiny. "Object-based focus involves a two-stage process that starts with the individualization of objects. Then, to connect object attributes, which are made accessible by feature maps, selective

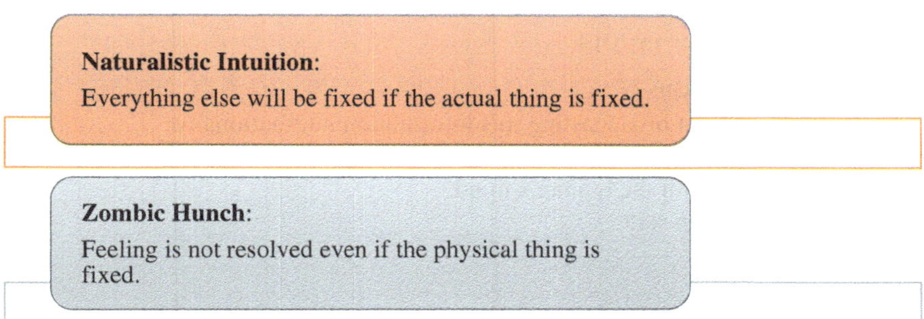

Fig. 3 The naturalistic intuition is that you fix everything else if you fix the actual problem, while the zombic hunch is that you somehow have not resolved the feeling even if you fix it physically

attention functions on these indexed objects, resulting in sustained object-based mental representations that enable object recognition" (Treisman and Gelade 1980).

As stated, attending to and accepting emotions must not be correlated with feeling them. However, there is also a fundamental social component of thoughts.

Technology has been progressing at an impressive pace, following developments in the mid-twentieth century, including the idea of Alan Turing for a general computing machine (Bauer et al. 2013).

This suggests that even understanding simple agencies in artificial intelligence programs is complicated. A similar but much more challenging topic concerns awareness. In terms of halting functions, artificial intelligence systems are defined. This experience of consciousness liberation from stopping roles is profoundly connected to the unification of human consciousness and the direction and actual importance of how feelings are perceived (Haladjian and Montemayor 2016). It is also strongly linked to the fact that consciousness does not appear vulnerable to duplication or reproducing, unlike programs. Like an attentional ritual, a function ceases, but consciousness never stops, nor does it rely on a program's conclusion (Mudrik et al. 2014).

Human intelligence also requires empathy, which creates a much more complex topic for artificial intelligence, as we have argued. And even if it were possible to apply the motor skills and step-by-step logic in robotics thoroughly, we would still not understand empathy in the artificial intelligence context. Artificial intelligence can, at best, generate specialized epistemic agents that can achieve unique objectives. However, artificial intelligence does not generate morally complex emotional agents with phenomenal interactions, motives, or emotions (Rinesi 2015; Riedl and Harrison 2015). It thus becomes ever more evident that while attentional mechanisms and even emotion-like reactions can be engineered, these artificial systems would not actually have a phenomenal perceptual experience.

Multiple Choice Questions

1. The test of "non-Turing" was introduced by:
   (a) Sigmund Freud in 1990.
   (b) Rene in 2002.
   (c) Ritchard in 1997.
   (d) Victor Argonov in 2014.

2. What makes us aware is:
   (a) The process of broadcasting this information internationally.
   (b) The consciousness.
   (c) The relativity of the human's mind.
   (d) The vegetarian state.

3. For many attention routines, the solid propensity for reducing these characteristics to mechanistic algorithmic routines can work, but not for feelings and agency, because:
   (a) The emotions are separated from attention.
   (b) Attending to and accepting emotions must not be correlated with feeling them.
   (c) Emotions are involuntary.
   (d) Robots have no emotions.
4. The capability to identify and communicate with other objects is necessary:
   (a) To enhance computer vision.
   (b) To build more interconnecting robots.
   (c) To do more research and experiments.
   (d) No need for that.

Answers

1. (d)
2. (a)
3. (b)
4. (a)

References

Alexiou A, Simou P, Alexiou F, Chatzichronis S, Ashraf GM. Social and ethical impact of advanced artificial and biological enhancements. Open Public Health J. 2020;13(1):62–8. https://doi.org/10.2174/1874944502013010062.

Argonov V. Experimental methods for unraveling the mind-body problem: the phenomenal judgment approach. J Mind Behav. 2014;35:51–70.

Bauer H, Veira J, Weig F. Moore's law: repeal or renewal? 2013. Retrieved from McKinsey & Company.

Ernst A, Bertrand JMF, Voltzenlogel V, Souchay C, Moulin CJA. The proust machine: what a public science event tells us about autobiographical memory and the five senses. Front Psychol. 2021;11:623910. https://doi.org/10.3389/fpsyg.2020.623910. Published 2021 Jan 20.

Haladjian HH, Montemayor C. Artificial consciousness and the consciousness-attention dissociation. Conscious Cogn. 2016;45:210–25.

Hildt E. Artificial intelligence: does consciousness matter? Front Psychol. 2019;10:1535.

Hromiak M. A new charter of ethics and rights of artificial consciousness in a human world. arXiv preprint. arXiv:2010.12019; 2020.

Koch C. Will machines ever become conscious? Scientific American; 2019.

Manzotti R, Chella A. Good old-fashioned artificial consciousness and the intermediate level fallacy. Front Robot AI. 2018;5:39.

Mudrik L, Faivre N, Koch C. Information integration without awareness. Trends Cogn Sci. 2014;18(9):488–96. https://doi.org/10.1016/j.tics.2014.04.009.

Oppy G, Dowe D. The Turing test. Stanford Encyclopedia of Philosophy (Spring 2011 Edition); 2011.

Riedl MO, Harrison B. Using stories to teach human values to artificial agents. Paper presented at the 2nd international workshop on AI, ethics, and society. 2015. http://www.aaai.org.

Rinesi M. The price of the internet of things will be a vague dread of a malicious world. IEET.org website. 2015. http://ieet.org/index.php/IEET/more/rinesi20150925.

Searle J, et al. Minds, brains, and programs. Behav Brain Sci. 1980;3(3):417–57.

Shieber S. The Turing test: verbal behavior as the hallmark of intelligence. Cambridge: MIT Press; 2004. ISBN 978-0-262-69293-9.

Treisman A, Gelade G. A feature-integration theory of attention. Cogn Psychol. 1980;12(1):97–136. https://doi.org/10.1016/0010-0285(80)90005-5.

Vervoort L, et al. Artificial consciousness, superintelligence and ethics in robotics: how to get there? Mind Matter. 2021;19(1):7–43.

Relativity of the Human Mind

Hashim Talib Hashim and Mustafa Ahmed Ramadhan

1 Introduction

Relativity means the matter or anything that will be discussed can differ with the situations and the circumstances like space and time. So people can do things and think it is right in some situations, but the same actions will be wrong in other circumstances. For example, a man can drink alcohol on certain occasions with controllable amount, but this action is wrong when it is done while driving because the situation is different here and the consequences are massive. Also, I can think that I am a child and whatever I do is to learn from it, but when I am an adult and do the same mistakes, its consequences will be unforgivable because the time is different and mind is different as well.

Humans' minds are dealing with the actions and giving orders in the same way of relativity so to keep a balance between our surroundings and our bodies and believes.

Any conflict or imbalance between our minds and our environment will create an issue that the mind cannot deal with, and this leads to psychological collapse.

In this chapter, we will discuss this topic from many fields and focus on a new presentation for the relativity theory in human consciousness as we believe that the human's consciousness is relative as well and we should deal with it depending on this idea.

H. T. Hashim (✉) · M. A. Ramadhan
University of Baghdad, College of Medicine, Baghdad, Iraq

H. T. Hashim, A. Alexiou (eds.), *The Psychology of Consciousness: Theory and Practice*, https://doi.org/10.1007/978-3-030-90692-4_10

2 The Absence of Ego

The lack of ego is an extreme limit to pro-activity and leaves people passively entering your path. Nowadays, we have the virtual path; it's quick but shallow even if the book was there before. Even for some cosmic reason, people cross our path, which has never changed. Today, we have a wide variety of networks. We can engage with crowds in Facebook groups, for instance, for the first time in human history, which makes me believe that conscious evolution can accelerate.

The "teal" campaign, as "green" is the end of life, has also become almost mainstream. However, if you are still not familiar with the color code of evolution and world views, please see my article on Spiral Dynamics Integral.

3 Conscious Evolution and the Future of Mankind

We may witness a significant transformation in our capacity to cope with our surroundings, living beings, and nature in general while we are there. The transformation will be dramatic, as crucial individuals will get a broader perspective and break free from the tiny circular circles of interest that have surrounded them up to this point (Styer 2011).

When they reach the higher levels, it is not necessary for anybody, or even a plurality of individuals, to acquire this degree of awareness for the connectivity of all to reoccur to a critical mass of people. As the global view expands, we will take more outstanding care of social justice, intentionally promote peace, clear up the trash in the oceans, replant trees, and so on, climbing the rungs of the spiral phases. This, by the way, would create a few jobs. We shall develop a healthier planet for all living beings to ensure a safe spiral of evolution.

4 Language Development and Consciousness

Language is part of human nature and being. There are compelling grounds to believe that no other entity possesses a distinctively human talent. Unlike other animals, we acquire complex linguistic norms rapidly and without purposeful, conscious effort in our early years. As our vocabulary expands, we may investigate the structure of phrases, the function and importance of terminology, and how phrases and words might be employed for social and practical goals (Psychology Today 2021).

Social actions become more complex as a result of language. Language improves reasoning skills since it allows us to express our views explicitly. Furthermore, language necessitates the habit of openly asking for explanations and reasons, which serves as a foundation for information acquisition. Homo sapiens became sapiens primarily as a result of language (Sorli et al. 2017).

The link is getting more difficult to establish because awareness appears to exist in many animals, but language does not. Language appears to have formed relatively lately in humans; awareness has evolved in many species, at least in its fundamental experiencing form, and maybe much earlier than language. Language and awareness appear to have a meaningful link, at least intuitively. Language may also be a necessary prerequisite for awareness. However, if language is only an exceptional human capacity, it will restrict human awareness; consequently, the issue's complexity must be addressed (Psychology Today 2021).

First, we need to understand language and its function in cognition. Thus, language can be an ability to speak and how thoughts are encoded, as the format for inferential reasoning.

5 Relativity of Human Behavior

Relativity in our theory means seeing things from different perspectives that change with situations. So, for example, when you see someone has a terrible situation, you will have your own opinion and behavior toward the issue. In contrast, the opinion and the behavior will differ when you are in the middle of the fire (Thierry 2016).

So, the relativity in psychology is not related to space and time; it relates to situations. When the situation differs, the behavior differs too. For example, when we put people in a bad situation like fighting and ask the other group to behave, they advise them to calm down and think clearly. When we make the same group fight and ask them to calm down, they respond aggressively. The same situation and the same people but different behavior is our relativity. The control factor in these situations is the mind. Your mind interacts with situations depending on the data it receives from the visual perspective with feelings and emotions. So, clearness during behavior can affect decisions and, thus, the results.

Multiple Choice Questions

1. The main interest of Tolman was:
   (a) How the brain can be relative to other systems.
   (b) The thinking process is managed by a behavioral process or cognitive method.
   (c) How neurons work.
   (d) How the relativity affects the mind.

2. The TCR actually advocates:
   (a) A promising approach to true artificial, particularly with artificial consciousness distinct from humans and animals.
   (b) The differences between the animal and human consciousness.
   (c) The electricity of the mind.
   (d) The robotic consciousness.

3. The cognitive path includes:
   (a) Searching for cognitive tasks related to conscious human experience and then checking how well those tasks are performed by the target animal species when the stimuli are presented in a specific modality.
   (b) The path to human consciousness by a robotic technique.
   (c) The cognitivity of the human mind.
   (d) The psychological path of the human.

4. .... is an extreme limit to pro-activity and leaves people passively entering your path.
   (a) Absence of Ego.
   (b) Absence of Superego.
   (c) Absence of Id.
   (d) All the above.

5. The relativity in psychology is:
   (a) Not related to space and time.
   (b) Not related to situations.
   (c) Related to both situations and space and time.
   (d) Not of the above.

Answers

1. (b)
2. (a)
3. (a)
4. (a)
5. (d)

References

Psychology Today. Consciousness and language. 2021 [cited 2021 Nov 13]. https://www.psychologytoday.com/us/blog/theory-consciousness/201608/consciousness-and-language.

Sorli A, et al. Advanced relativity: multidimensionality of consciousness and mind, origin of life, psi phenomena. NeuroQuantology. 2017;15(2):109–17.

Styer DF. Relativity for the questioning mind. Baltimore: JHU Press; 2011.

Thierry G. Neurolinguistic relativity: how language flexes human perception and cognition. Lang Learn. 2016;66(3):690–713.

Death and Consciousness

Hashim Talib Hashim and Mustafa Ahmed Ramadhan

1 Introduction

In terms of consciousness, it is possible to describe existence.

Death is more of a mechanism in some societies than a single occurrence. Therefore, a gradual change from one spiritual state to another is inferred (Parnia 2014).

Eternal oblivion is also called the belief in the irreversible loss of consciousness after death. The word afterlife describes the assumption that the stream of knowledge is maintained after physical death. Neither is likely ever to be proven without the pondered having to die (Parnia 2014).

Detritivores, species that recycle detritus, may further decompose organic material, returning it to the atmosphere for reuse in the food chain. These chemicals ultimately become absorbed and assimilated into a living organism's cells. Earthworms, woodlice, and dung beetles are some examples of detritivores (Parnia 2014).

2 The Difference Between Consciousness and Death

Death is not the opposite of consciousness. It is the lack of consciousness—that is missing. In practical terms, the fractal catalytic model assumes that a macroscopic coherent wave function—a coherent matter-wave—supports the neural activity. The neural association of consciousness is this macroscopic quantum coherent wave function. Once this wave function collapses, the topic will do so as well.

H. T. Hashim (✉) · M. A. Ramadhan
University of Baghdad, College of Medicine, Baghdad, Iraq

H. T. Hashim, A. Alexiou (eds.), *The Psychology of Consciousness: Theory and Practice*, https://doi.org/10.1007/978-3-030-90692-4_11

With plenty of parallels and dislikenesses, rest and demise are if you are in yoga, one school demonstrates to you that contemplation is a specialty of kicking the bucket (actually what it means is that we totally free body consciousness as reflection happens to us), cognition achieves a past state psyche, but in any known dialect, this cannot be reflected in words. As troublesome as the sweet portrayal of any dialect (Lund 2016).

A state connected with the body and the brain is added to your inquiry about when one more condition of a presence called otherworldly is included. Profound rest is near-death as consciousness is permitted to choose whether to backpedal to mind and body or free itself and choose another desire to be conceived again. Cognizance joins itself as it leaves the body with the last thought in the brain.

Mercifully take note that cognizance will relate to only one thing at any given moment. It gets pulled in to alter, too. So those out of the body will lose the progressions realized by the gross detects, hence no more improvements in this direction, the last idea is clutched.

Advancing fixation and culminating reflection help us, considering the fact that the consciousness comes to the past brain, to use the embodiment of the gross faculties placed away in the memory. I used pith and not the gross detect feature. The dialect is not vital along these lines.

3 Near-Death Experiences

It is defined as the moments that come during some life events prior to death or when the individual is about to lose his or her life (Koch 2021).

These concrete and fantastic encounters have the potential to alter their lives permanently. NDEs are not fanciful imaginings (Van Lommel 2006).

The media coverage of near-death experiences (NDEs) is likely to have led to expectations about how people could feel following such experiences (Carter 2010).

Local brain areas fall dark one after the other, like a municipality losing power one community at a time.

Given the power interruptions, this experience has the potential to yield the bizarre and unusual accounts that comprise the corpus of NDE reports. The NDE is as genuine to the person experiencing it as anything manufactured by the imagination during regular waking hours. The mind and consciousness are extinguished when the whole brain shuts down due to total power loss (Klein 2020; Koch 2021).

Again, an aura, a special feeling unique to a patient that indicates an impending assault, will precede them.

Following a seizure, you may experience changes in the apparent proportions of items, strange tastes, scents, or physical sensations, déjà vu, depersonalization, or euphoric sentiments.

4 Consciousness After Clinical Death

Many beliefs can lead us to the point that the consciousness may remain to exist after death or at least remains working (Van Lommel 2010).

Pharmacological manipulation is used by various medications that influence neural activity by interfering with neurotransmission, resulting in vision, emotion, awareness, cognition, and behavior changes. Psychoactive drugs are classified into four groups based on their pharmacological effects: euphoriants, stimulants that induce transient improvements in either mental or physical functions, depressants that depress or decrease arousal or stimulation, and hallucinogens that may cause hallucinations, anomalies in perception, and other significant subjective changes in perceptions, emotions, and consciousness (Metcalf and Huntington 2014).

Death was formerly defined as the cessation of heartbeat and breathing (cardiac arrest). Nonetheless, the development of CPR and quick defibrillation has rendered this paradigm obsolete, as it is frequently feasible to resume breathing and heartbeat.

Life may often be preserved without a functional heart or lungs using various life support systems, organ transplants, and artificial pacemakers; events that used to be causally associated with death no longer kill under all conditions (Metcalf and Huntington 2014).

Multiple Choice Questions

1. What is death?
   (a) Death is the opposite of consciousness.
   (b) It is the continuity of life.
   (c) It is the freedom of the soul.
   (d) It is the lack of consciousness.

2. Heart attack can be considered:
   (a) A near-death experiences.
   (b) A vegetarian state.
   (c) An unconscious state.
   (d) The start of the death.

3. Putting the person inside an MRI machine or surrounding his head with electrodes:
   (a) He reaches the NDE.
   (b) HE does not reach the NDE.
   (c) It is a start for the NDE.
   (d) It has no relation with NDE.

4. …………….. is the irreversible loss of consciousness after death:
   (a) Near-death experience.
   (b) Consciousness.
   (c) Eternal oblivion.
   (d) Death.

Answers

1. (d)
2. (a)
3. (b)
4. (c)

References

Carter C. Science and the near-death experience: how consciousness survives death. Toronto: Simon and Schuster; 2010.

Klein A. The death of consciousness? James's case against psychological unobservables. J Hist Philos. 2020;58(2):293–323.

Koch C. What near-death experiences reveal about the brain. Sci Am. 2021 [cited 2021 Nov 13]. https://www.scientificamerican.com/article/what-near-death-experiences-reveal-about-the-brain/

Lund DH. Death and consciousness. Jefferson: McFarland; 2016.

Metcalf P, Huntington R. Celebrations of death. Cambridge: Cambridge University Press; 2014.

Parnia S. Death and consciousness—an overview of the mental and cognitive experience of death. Ann N Y Acad Sci. 2014;1330(1):75–93.

Van Lommel P. Near-death experience, consciousness, and the brain: a new concept about the continuity of our consciousness based on recent scientific research on near-death experience in survivors of cardiac arrest. World Futures. 2006;62(1–2):134–51.

Van Lommel P. Consciousness beyond life. New York: HarperCollins; 2010.

Index

A
Abuse, 89
Addiction, 82
Alcohol, 83
Altered states of consciousness, 99, 101
Altruism, 76
American Psychiatric Association (ASA), 58
Amphetamine, 81
Amygdala, 19
Analgesics, 86
Antisocial personality disorder (ASPD), 70
Artificial consciousness
  attention, 123
  behavioral processes, 122
  consciousness liberation, 124
  generality, 122
  human intelligence, 124
  Lagom, 122
  mind processing capabilities, 121
  naturalistic intuition, 123
  object-based focus, 123
  strong consciousness, 121
  subsidiary mechanisms, 121
  tools, 121
Artificial intelligence, 119, 124
Attention, 112, 114
Automatic processes, 111
Autosuggestion, 32
Avoidant personality disorder (AVPD), 73

B
Barbiturates, 84, 85, 92
Benign sleep myoclonus of infancy (BSMI), 48
Benzedrine, 81
Biological influences, 104, 105
Borderline Personality Disorder (BPD), 71
Brain–mind relationship
  consciousness, 21
  neuroanatomy, 21
  senses, 21
Brain regions
  left and right hemispheres, 19
  limbic system, 19
  memory, 20
  neocortex, 19
  pathologies, 19
  sensory cortex, 21

C
Cataplexy, 45
Central sleep apnea (CSA), 43, 44
Choice blindness, 100
Circadian, 39
Circadian rhythm sleep-wake disorders, 43
Clinical death, 133
CNS depressants
  alcohol, 83
  barbiturates, 84
CNS stimulants, 92
  amphetamine, 81, 82
  caffeine, 81, 82
  cocaine, 83
  nicotine, 81
Cocaine, 83
Codeine, 86
Cognitive-behavioral therapy (CBT), 73, 74
Cognitive perception, 110

H. T. Hashim, A. Alexiou (eds.), *The Psychology of Consciousness: Theory and Practice*, https://doi.org/10.1007/978-3-030-90692-4

Coma, 22
Computer intelligence, 119
Computer programs, 123
Computer science, 119, 122
Confusion, 23
Confusional arousal, 54
Connecting memories, 29
Conscious evolution, 128
Consciousness, 51, 85, 109, 120
  abnormal states, 22
  agreement, 10
  awareness, 6
  death, 131
  decisions and actions, 13
  definition, 2
  development, 1–3
  developmental psychology, 10, 11
  domains, 4
  explanations, 1
  Freud, S., 6
  language, 128, 129
  levels, 8
    ego, 27
    Id and ego, 27
    personality, 27
    preconscious, 29
    Superego, 27
  medicine, 22
  meditation, 4
  memory, 132
  mind and body interactions, 4
  neuropsychological research, 13
  neuropsychology, 14, 15
  neuroscience, 14
  object, 13
  personal ID, 2
  psychiatry, 22
  social psychology, 11
Consciousness awareness technologies, 121

D
Daydreaming, 100
Death, 133
Defense mechanism
  classification, 58, 59
  definition, 57
  immature defenses, 60, 61
  mature defenses, 65, 66
  mental health, 57
  moral values and principles, 57
  neurotic defenses, 62–64
  pathological defenses, 59
Delete Button, 31
Delirium, 24
Dependence, 84
Detritivores, 131
Dexedrine, 81
Dialectical behavioral therapy (DBT), 71, 74
Digital consciousness, 119
Divided-consciousness theory, 114
Dopamine agonist, 92
Dreams
  conscious mind, 49
  consciousness, 51, 52, 54
  definition, 49
  lucid, 51
  meta-awareness, 51, 52
  neurobiological perspective, 50
  postconsciousness, 52
  properties, 49, 50
  self-awareness, 52
Dual processing, 97
  change deafness, 99, 100
  choice blindness, 99, 100
  selective attention, 97, 98
  two-track mind, 97, 98
Dualism, 6

E
Electromyography (EMG), 49
Emotions control communication, 30

F
Fentanyl, 87
Freudian theory, 6, 7, 15

G
GABA agonist, 85, 92
Gibson's theory, 9
Glasgow Coma Scale (GCS), 23
Grady Coma Scale (GCS), 23

H
Habenula, 20
Hallucination
  LSD, 91
  marijuana, 89, 90
  MDMA, 90, 91
Histrionic personality disorder (HPD), 72
Human
  vs. machine consciousness, 120
Human awareness, 120
Human behavior

relativity, 129
Human complex conscious abilities, 14
Human consciousness, 5, 122
Human intelligence, 124
Human mind
ego, 128
human's consciousness, 127
relativity, 127
social justice, 128
Hypercapnic type, 44
Hypnosis, 102, 105
calming experience, 111
clinical and psychological applications, 114
hypnotherapy, 115
pain management and cancer supporter, 115
prevention of irritable bowel syndrome, 115
consciousness, 109
as divided consciousness, 113–114
emotional, 112
experiences, 113
hypnotic hypermnesia, 111
hypnotizable estimation, 110
hypnotized people, 109
induction process, 110
interactions, 110
and meditation, 112
as meditative condition, 111
multifaceted essence, 110
posthypnotic suggestion, 111
radical neurological improvement, 111
social interaction, 112
uses, 114
Hypnotherapy, 115
Hypnotic induction, 111
Hypocapnic type, 44
Hypothalamus, 23

I
Idiopathic hypersomnia (IH), 45
Implicit memory, 22
Inattention, 99, 101
Inattentional blindness, 99
Insomnia, 43
Intellectualization, 76
Irritable bowel syndrome, 115

J
Jet lag disorder, 53

L
Language development, 128
Laterodorsal tegmentum (LDT), 50
Lethargy, 22, 23
Locke's theory, 3

M
Machine consciousness, 119
MAOIs selective serotonin reuptake inhibitors (SSRIs), 88
Marijuana, 90
effects of tetrahydrocannabinol, 90
hallucination, 89
medical uses, 90
positive and negative effects, 90
properties, 90
Meditations, 4, 102–105
Mentalization-based therapy (MBT), 71
Meperidine, 88
Metacognition-oriented therapy, 68
Metacognitive interpersonal therapy (MIT), 74
Methadone, 88
Methamphetamine, 82
Morphine, 85–87

N
Narcissistic personality disorder (NPD), 71, 75
Narcotics
addiction, 85
codeine, 86
fentanyl, 87
meperidine, 88
methadone, 88
morphine, 85, 86
oxycodone, 89
Naturalistic intuition, 123
Near-death experiences (NDEs), 132
power interruptions, 132
Neurobiological theory, 54
Neuropsychology, 16
Nightmare disorder, 54
Non-human artificial consciousness, 122
Non-Rapid Eye Movement (NREM), 40

O
Obsessive-compulsive personality disorder (OCPD), 74
Obstructive sleep apnea (OSA), 43
Operative functioning, 30
Opioids, 86–88, 92
Oxycodone, 89

P
Paranoid personality disorder, 75
Parasomnia
  definition, 45
  dissociative disorder, 47
  hallucinations, 47
  NREM sleep
    confusional arousal, 46
    eating disorder, 46
    sleep terrors, 46
    sleepwalking, 46
  REM, 47
Pedunculopontine tegmentum (PPT), 50
Perception, 97, 102
Periodic Limb Movement Disorder (PLMD), 48, 54
Personal identity theory, 2
Personality disorders
  cluster
    ASPD, 70, 71
    AVPD, 73
    BPD, 71
    DPD, 73
    HPD, 72
    NPD, 71
    OCPD, 74
    PPD, 68
    SPD, 69, 70
    SzPD, 69
  definition, 67
  DSM-5, 67
Pharmacological manipulation, 133
Physical thing, 4
Postconscious, 15, 52
Posthypnotic amnesia, 111
Preconscious, 29
  adults, 29
  memories, 29
  psychiatry and psychology, 29
  thinking, 29
  words presentations are memory, 30
Psychiatric disorders, 66
Psychological influences, 104, 105
Psychostimulant, 82

R
Rapid eye movement (REM), 40
Recognition, 97
Refocusing attention, 102
Relativity, 127, 129
REM sleep behavior disorder (RBD), 46
Restless leg syndrome (RLS), 47, 53, 54
Robotics, 124
Robots, 119, 120

S
Schizotypal personality disorder (SPD), 69
Selective attention, 97–106
Sleep
  circadian, 39
  definition, 39
  disorders
    breathing, 43
    circadian rhythm sleep-wake, 43
    CSA, 44
    insomnia, 42, 43
    OSA, 44
  hypersomnolence, 45
  hypothalamus, 39
  NREM
    definition, 40
    heart rate, 42
    SWS, 42
    wakefulness state, 41
  parasomnia, 45
  REM, 41, 42
  SCN, 40
  types, 40
Sleep-related movement disorders
  bruxism, 48
  BSMI, 48
  leg cramps, 48
  PLMD, 48
  PSM, 49
  RLS, 47
  SRRMD, 48
Sleep-Related Rhythmic Movement Disorder (SRRMD), 48
Slow-wave sleep (SWS), 42
Social influence theory, 114
Social psychological investigation, 12
Social psychology, 12
Spinoza, 6, 16
Subconscious
  desire, 31
  emotions control communication, 30
  fear and negative talking, 30
  mind, 30, 32
  unconscious, 32

Subjective perception, 111
Subliminal perception, 32
Subliminal stimulation, 36
Superego, 7
Suprachiasmatic nucleus, 53

T
Transference-focused psychotherapy (TFP), 71
Turing test, 119

U
Unsatisfactory cognitive dissonance, 123

V
Vermis, 19

W
Weak artificial intelligence, 120–122

Z
Zombic hunch, 123

The manufacturer's authorised representative in the EU is Springer Nature Customer Service Centre GmbH, Europaplatz 3, 69115 Heidelberg, Germany. If you have any concerns regarding our products, please contact ProductSafety@springernature.com

Printed and bound by CPI Group (UK) Ltd, Croydon, CR0 4YY
30/09/2026
02225465-0012